Testimonials

"Bob's insights into disruptive change and how it might transform the insurance industry are visionary. At a time when every leading service provider is attempting to understand their customer's needs more clearly and answer the question is my value proposition compelling, Bob has gone further and designed a new insurance paradigm with the customer's needs at the center of the blueprint. The Performance Based Risk Management (PRBM) approach described in Appendix A has the potential to rewrite the value proposition in the insurance industry."

TONY ALLEN
MANAGING PARTNER
THE IMPACT GROUP

"I have known for many years that Bob Phelan is the best Risk Advisor in the construction business..... now I know why."

BILL DUCCI
CHAIRMAN/CEO
DUCCI ELECTRICAL CONTRACTORS, INC.
#300 2008 ENR SPECIALTY CONTRACTORS

"A broad-based approach to Enterprise Risk Management, which goes far beyond a review of minimal business insurance coverage, is already being embraced as a best practice by many of America's largest corporations. Likewise, moving away from the billable hour to a more value or performance based billing approach is already being employed by some of the nation's leading law firms and cutting edge corporate legal departments. I applaud Bob Phelan for advocating that these two concepts be merged and applied to his clients and others in the construction industry."

WILLIAM J. CASAZZA
SENIOR VICE PRESIDENT AND GENERAL COUNSEL
FORTUNE 100 COMPANY

"Mr. Phelan provides a very insightful look at construction risk management that should be mandatory reading for any stakeholder in a construction business. His "all in" metaphor aptly describes the magnitude of risk faced by most construction companies today and, when juxtaposed to the all too common practice of buying the cheapest insurance from any broker who can find it, illustrates the ineffectiveness of the "business as usual" approach to managing risk. For example, as respects contractual risk transfer, his observation that he has never seen compliance with the risk and insurance requirements in a construction contract should give pause. His experience is similar to mine – if there is any doubt about this issue, one need only to note the disproportionate amount of litigation regarding the scope of insurance afforded an additional insured. In sum, Mr. Phelan's book clearly explains the two ends of the spectrum - the insurance broker as "order taker" and the insurance broker as "risk advisor," including the crucial implications of each and the value proposition offered by his organization. Which insurance broker or "left tackle" would your organization rather have protecting your blind side?"

CRAIG F. STANOVICH, CPCU, CIC, AU
PRINCIPAL & CONSULTANT
AUSTIN & STANOVICH RISK MANAGERS LLC

"Bob Phelan has been providing expert risk management services to O&G Industries for over 20 years. His practical insights and innovative thinking in dealing with unique exposures and changing conditions have made a significant bottom-line impact on our company. If you are looking for a trusted advisor to protect both your company's blind side and more readily apparent risks, his expertise and commitment to the client will serve you well."

KENNETH W. MERZ
CORPORATE SECRETARY
O&G INDUSTRIES, INC.
#157 2008 ENR 400

"Controlling your Total Cost of Risk is crucial in today's market-place and economy. In the average middle market company, your pure insurance premiums represent less than 50% of your Total Cost of Risk. Bob Phelan has mastered the art of designing and implementing Risk Reducing Plans that allow his clients to control their TCOR while becoming a more attractive risk to the insurance carriers."

ROGER SITKINS
FOUNDER/CEO/HEAD COACH
SITKINS GROUP, INC.

"Bob Phelan and the Litchfield Insurance Group have analyzed, exposed, and tackled the potentially destructive flaws in the risk management and insurance buying process of construction companies. Their systems and processes are absolutely essential to the sustainability and growth of construction companies in an increasingly complex, litigious, and competitive world. Construction companies must at least listen to, but preferably adopt, The Litchfield Insurance Group's method of managing and leveraging risk. Of course, as Dr. W. Edwards Deming said, 'No one has to change. Survival is optional.'"

FRANK PENNACHIO
CO-FOUNDER AND DIRECTOR OF LEARNING
INSTITUTE OF WORK COMP PROFESSIONALS

"In Broke, Bob Phelan, President and CEO of The Litchfield Insurance Group crystallized his years of insurance experience into a razor-sharp book that turns conventional insurance ideas and risk reduction management on its head. This book should be essential reading for all construction industry business owners that are looking to strengthen their businesses and boost their profits well into the future."

CHARLES ANDERSON
PRESIDENT, SELLING SKILLS INSTITUTE
AUTHOR OF: THE SECRET TO SALES GREATNESS

"A straightforward universally applicable method for managing risks of small to mid-size construction companies. A must read for those owners/CEO's interested in protecting their most valuable asset - their company."

TODD BATESON
SENIOR EXECUTIVE
FORTUNE 100 COMPANY

"The principles within "Broke..." have been a cornerstone of our business for many years thanks to our long-valued relationship with Bob and his firm. Though none of us can predict what the future holds, these fundamentals have provided for us a means of assessing potential future risks and safe-guarding against them, while simultaneously allowing us to focus on doing what we do best."

RICHARD J. DUCCI
PRESIDENT/CFO
DUCCI ELECTRICAL CONTRACTORS, INC.

"I think the book is informative and insightful. It's a great read not only for those who contractors who need more specialized risk management advice but also a great refresher for those with extensive risk management training. There are theories in this book (protecting one's blind side and understanding the total cost of risk to name just two) that are vital to the growth, protection and success of all businesses. Contractors will be thankful and in a better place having read this book and implementing the ideas and strategies outlined."

RICHARD GROSS
2ND VICE PRESIDENT
MAJOR DOMESTIC REINSURANCE COMPANY

"Broke: The Broken Contractors Insurance System and How To Fix It is an attempt to change the way risk is viewed and quantified, the way insurance has historically been purchased and the benefits a risk advisor provides to mitigate the total cost of risk. It provides an insightful look into the insurance mechanism, process and distribution by giving illustrations from the view of the insured, agent and insurance company. Most insurance claim reports begin with the sentence, "I didn't see it coming". This book prepares the construction company executive and reduces the number of surprises in the clients risk profile."

ROBERT J. BROOMALL, CPCU

BROKE

BROKE

The broken contractor's insurance system and how to fix it

Robert Phelan, ARM, CRIS

Published by Advantage, Charleston, South Carolina.
Member of Advantage Media Group.

ADVANTAGE is a registered trademark and the Advantage colophon is a trademark of Advantage Media Group, Inc.

Printed in the United States of America.

ISBN: 978-1-59932-106-6
LCCN: 2009924240

This publication is designed to provide accurate and authoritative information in regard to the subject matter covered. It is sold with the understanding that the publisher is not engaged in rendering legal, accounting, or other professional services. If legal advice or other expert assistance is required, the services of a competent professional person should be sought.

Most Advantage Media Group titles are available at special quantity discounts for bulk purchases for sales promotions, premiums, fundraising, and educational use. Special versions or book excerpts can also be created to fit specific needs.

For more information, please write: Special Markets, Advantage Media Group, P.O. Box 272, Charleston, SC 29402 or call 1.866.775.1696.

Visit us online at **advantagefamily**.com

To my lovely wife and wonderful children. As the people I cherish most in the world, I will do my utmost to protect your blind side, now and forever.

Table of Contents

Acknowledgements

To my father who introduced me to the wonderful world of insurance and risk management and my mother for her unwavering belief in my potential.

To Dan Kraut for our friendship, our business partnership and for giving me the opportunity to learn and experiment with the ideas which form the basis of this book.

To Steve Ward. Without your belief in me and my organization, this book would not have been possible.

To Julie Roads my tireless editor. You made this project a whole lot easier than it could have been for a first time author.

To Dan Kennedy, Bill Glaser and Lee Milteer, my coaches in our Peak Performers Mastermind Group. All of you have been a great inspiration to me. You gave me the courage to begin this project and I am forever grateful.

To Adam Witty, my publisher, and his staff at Advantage Media. Thanks for making it so easy for me to write this book.

To the risk takers and innovators at Valorem Law Group. Without your inspiration, I wouldn't have had the courage to take the bold steps outlined in this book.

To my long time client and great friend, Bill Ducci. Thank you for challenging me and supporting me is so many ways over our 30+ year relationship. Your loyalty and friendship mean more than you will ever know.

To Ken Merz, Rich Hall and their associates at O&G Industries for believing that our company has the capability to manage their risks as well as any national broker.

Thanks to Ray, Greg, David, Bob and Rod Oneglia for your loyalty to our long term relationship.

To my coach, mentor and friend, Roger Sitkins and his team at Sitkins International. He has been my greatest resource in developing the great capabilities our company has today.

To Ron Jodice for believing in the concept of Performance-Based Risk Management™. He took a chance on a new idea and gave me the courage to expand its possibilities.

To Charlie Anderson, my good friend and coach. His wise counsel and patience have been a determining factor in my company's success.

1 The Genius of the Left Tackle

Most small- and medium-size businesses in America are family businesses, oftentimes multigenerational. These are the types of clients I work with every day. Over the years of building that business, many family members have put in their blood, sweat, and tears to build more than a company, they've built a foundational part of their lives – and it's the most valuable asset they own in the world. Far beyond the value of their retirement plan, far beyond the value of their house, their business and its success and longevity is critical to their existence. It's a bedrock that provides life-support and income for many generations, and it's an asset that must be protected.

Shockingly, I've discovered during my professional career in insurance and risk management that these business owners, especially the construction industry business owners to whom this book is addressed, don't often realize the enormity of value their asset possesses – for both themselves and their future generations. And if they do, they aren't taking the steps to protect it.

There seems to be a misconception on the part of these business owners. They tend to project their past experience onto the future when it comes to risk. The mentality is that the insurance they've always had (based on standard practices, lowest cost, agent recommendation, and industry norms) will work for them into eternity.

But particularly with middle market construction companies, you can't project the past into the future. Prudent risk management revolves around catastrophic loss, what we call *severity* in the business. A business owner shouldn't pay as much attention to the everyday things that insurance and risk-management protect against, such as a fender bender that costs twenty-five hundred dollars – the types of incidents covered under common insurance packages. Owners should be paying more attention to the single, large event that can destroy their business. These are the events that risk management professionals should be working to proactively prevent for the company.

Because of the typical multigenerational nature of the family construction business, there's a lot at stake; and because there's a lot at stake, business owners need to look at a bigger picture when they think about risk management. As someone who has been a practitioner in the field of insurance and risk management for more than thirty years, I see risk through a much bigger lens. Insurance brokers and risk management professionals see what's happening to business owners of all types and sizes over an entire state, region, or even the world. Disaster happens to businesses all the time. I see quick blurbs in the newspaper or on the news at night almost every day where something horrible has happened to a local, regional or national company. But what isn't seen in that news clip is how in only a moment that business, often a family business, was suddenly destroyed in its 80th, 100th, or180[th] year.

When I see these stories, it is all too clear to me. I have seen construction businesses destroyed in an instant by something that they didn't anticipate and weren't prepared for. The purpose of this book is to help you, the business owner, prepare for those unexpected events by putting a risk management program in place that will actively protect your business for this generation and the next.

There's a distinction between an insurance agent/broker and what I call a "Risk Advisor," which I will go into in more detail in Chapter 5.

For now, suffice it to say that a Risk Advisor should be a key part of what I call your "trusted advisor network" – a network that includes your attorney, CPA, banker, and other types of consultants that are critical to the operation of your business. The Risk Advisor looks at every operational and financial component of your company, taking a very broad view of your business and what can happen, and brings a unique perspective that is far beyond looking at just insurance policies and coverage. Your Risk Advisor should be in a position to predict not the future but the likelihood of the catastrophic things that could happen to threaten your business and the areas where you really need to focus your time, attention, and resources for the prevention of such a destructive event.

My experience has shown me that the Risk Advisor, if competent, trained, experienced, creative, and full of wisdom, should be the single most important advisor that a medium-size construction company has. The mission of this book is to help the construction industry – an industry that is largely comprised of multigenerational, or soon to be, multigenerational companies – understand the impact, necessity, power, and reach of risk management on their company's present...and their future.

The Blind Side

There's a book that was written recently called *The Blind Side*. It's about the history of football and how football strategy has changed over the last twenty years. I'm from New England where the New York Giants have played a prominent role during the last quarter century. It all began when Lawrence Taylor (or "LT" as he came to be known) began playing for them in the early 1980's. Taylor was a right linebacker, a feared defenseman who had quarterbacks shaking in their cleats. He famously stated his mission as destroying, or even, killing the quarterback – and it all came to life when he broke two bones in Joe Theismann's leg in that famous Redskins game.

Most quarterbacks are right-handed and their blind side is the left. When they turn their heads to follow that right arm, they can't see what's coming at them from the other side—and if it happens to be Lawrence Taylor fast approaching, then they need the best protection they can possibly have in the left tackle position.

Literally, because of LT, football took on a new dimension, and the left tackle position on the offensive line suddenly became one of the most important positions on the team.

Simultaneously, Bill Walsh, the famous coach of the San Francisco 49ers, built an offensive strategy based on short passes. Showcased best by Joe Montana and his successor, Steve Young, they were able to move the ball down the field and win three Super Bowl Championships with the new strategy. The key element was the ability of the left tackle to protect the blind side of the quarterback.

The story of Lawrence Taylor, the left tackle and the new offensive strategy correlates directly to your company. Why? Because you damn well better know the answer to this question: **"Who is your left tackle?"**

Everyday you get up, you bid on jobs, you try to execute on the contracts that you have in place. You're building things. You try to make money. You're doing all the things that a construction company owner does, but you have a blind side that contains all those risks that can potentially destroy your business. *The risks that exist in our world today are on a much larger scale than they've ever been before.*

Early in 2008, I happened to play in a celebrity golf tournament in Connecticut called "The Ahmad Rashad Celebrity Classic" to benefit the Boys and Girls Club of Hartford, Connecticut. Lo and behold, one of the players there was Lawrence Taylor.

As we played golf that day, there were leader boards staggered around the course showing who was winning. I happened to be playing with Ahmad Rashad, and we were playing pretty well. Our team was zigging and zagging and putting a good number up on the board, but there was no way we could get past Lawrence Taylor, who held the lead most of the day. Later in the tournament, I heard Ahmad Rashad, who is a pretty big guy, talking to some other really big guys, football and baseball players, questioning whether Lawrence Taylor really could have had such a low score. They all said to each other, "Well, you go ask him." The others would say, "No, you ask him. You do it." No one wanted to confront LT. Case in point, Lawrence Taylor was a mean and fierce competitor – quarterbacks had a reason to be up at night staring at the ceiling wondering what might happen to them if he got through that offensive line on Sunday afternoon.

As a business owner, do you ever go to sleep at night staring up at the ceiling, wondering what's going to get through your offensive line? There's a world out there full of Lawrence Taylors ready to destroy your business. I'm guessing that as you look out at your construction operations, you see heavy equipment, you see workers exposed to extreme and stressful work environments, you see large fleets of automobiles and trucks on the highway every day – subject to weather conditions and other drivers. All of those things look like potential accidents waiting to happen, full of what ifs.

In 2001, a salesperson in a pickup truck, working for a building supply wholesaler, was talking on his cell phone. He was inattentive, ran a stop sign, and plowed into a vehicle, injuring a seventy-eight year-old passenger. She sued the owner of that pickup truck, which was a large multi-state company, and the jury ultimately awarded her **twenty-one million dollars**. I think that is a frightening statistic for any company to see because probably all of us have gotten distracted while driving, whether it's by a cell phone or something else. It just takes one moment, one quick point of distraction, and catastrophe wipes out a company built and sustained for generations.

I had an experience with a client earlier this year in which a job site crew had just finished a safety meeting when the foreman, who was sitting in a loader and wasn't paying attention, dropped the bucket on another worker's leg and just about caused an amputation. If it weren't for the quick response and medical attention he received, this worker probably would have lost the leg and it would have been a multimillion-dollar loss instead of a quick recovery.

It only takes a moment. It doesn't take more than one multi-million dollar catastrophe to either leave a business with an uninsured loss or in a position with such exorbitant insurance premiums that their budgets don't allow for them to compete with their competitors.

Force yourself to take a reality check and ask this question, "if one of your workers were involved in an automobile accident where someone was seriously hurt as a result of talking on their cell phone, could you afford a twenty-million-dollar jury verdict?"

Another sobering example occurred several years ago when a bridge contractor had an employee working in an elevator shaft. It was their most experienced and safest employee, and it was the end of the shift. This employee thought to himself, *I just need to do one more task, take one more minute*, and he stretched himself beyond the safe zone and fell thirty or forty feet down the elevator shaft, landing on his back and his legs. The worker would have been killed if there hadn't been a bunch of loose pieces of plywood stacked in the bottom. As it was, he broke the femurs in both legs but wasn't killed.

As a result of that accident, the Experience Mod Rate of this contractor went through the roof. The insurance premium went through the roof because all of a sudden, the underwriting community could see in vivid detail just how much money a claim with this contractor could cost them, so no one wanted to write their insurance. It put this construction company on the edge of viability, and it all only took a moment.

An Experience Mod Rate is something that you are all probably familiar with relative to workers' compensation. It's a number that compares your workers' compensation claim experience to other contractors that do the same thing. That mod rate is 1.00 if you're average with everybody else. It's less than that if you're better. It's more than that if you're worse. Every contractor's goal should be to get their Experience Mod Rate, or EMR, to the lowest point possible. Later in the book, we'll talk about some of the ways that we can help contractors accomplish that.

You don't always get what you pay for

When my firm talks with prospective clients, two of the things we focus on are 1) What you get for your insurance premium, and 2) What you get for the commission paid to your insurance agent or insurance broker.

Ours is an industry that hasn't evolved very much in the thirty plus years I've been practicing. It's still an industry in which a company pays a premium to an insurance company and then an insurance company pays a commission to the agent or broker who placed that business. It's rare that a contractor looks at those numbers and says, "What am I getting for that money? What's my return on that investment?" Our industry is still way too transactional and we're going through a period right now when Risk Advisors, or the better insurance brokers, are starting to offer a different value proposition than what's been offered in the past.

Traditionally, insurance agents and brokers are very transactional. They're selling insurance policies, and then, servicing those insurance policies. But if you're just buying insurance and treating it as a transaction, you're not looking at that bigger picture, let alone your blind side. You're not looking at the value of that asset and all the different risks that threaten it. In today's world, construction company owners need someone who offers much, much more than the least expensive policy and reactive service. They need someone who offers a much richer value proposition, someone who can look holistically, creatively, and proactively at their business; someone who can see all of the different risks. Today's construction company owners need more than insurance, they need a trusted advisor that will help them protect their most valuable asset.

As our world and its complexity grow, the blind side of a business gets bigger and bigger at an exponential rate. Later in this book, we'll explore compensation, which I think is long overdue for dramatic change. Insurance brokers and agents shouldn't be paid standard commissions by insurance carriers. They should be paid a fee agreed upon by their client for the work, advice, and counsel they're going to provide. If you really want someone to be your left tackle, to protect your blind side, maybe it costs more money than you're used to paying in the form of a commission; but in return for that, you're going to receive a commitment from a Risk Advisor that the blind side is going to be protected. There are going to be guarantees in terms of performance, the compensation will be equal to the performance. There's going to be a larger team of professionals that protect you in many ways beyond simply insurance policies.

Think about it this way. Somebody said in a seminar that I attended recently that the two biggest sections of the phonebook are attorneys and pizza parlors. And as I've traveled around the country, from time to time when I'm in a business or in a hotel room and I see phone book, I'm reminded to check that theory. It's not that I have anything against attorneys; attorneys provide a very valuable service to all of us. I use attorneys all the time. But there are lots of attorneys out there on the personal injury side who are after your blind side, when your blind side is left unprotected, they benefit.

I happened to be in Charleston, South Carolina, and the phone book had just arrived at the office I was visiting. I decided to test the theory, and I looked inside. As you typically find these days, there was an attorney on the inside cover. In the Charleston phone book, the attorney section runs from page 50 to page 125. It's 75 pages long! The whole phone book is only 773 pages. So ten percent of the phone book is attorneys – ten percent!

I asked a couple of my business associates who were in the room with me how many pages they thought attorneys took up in the phone book. One guess was eight and the other was twenty. So needless to say, most of us don't realize how big the legal profession is and how well they advertise. It isn't just the phone book, it's on the sides of busses and overpasses and billboards. Personal injury is a big business, and it's probably your biggest blind side.

This might shock you, but my company has come up with a philosophy in which we don't believe in insurance brokers competitively bidding for clients based on the cost of the insurance. Yes, it's traditional, it's the industry norm, but it has created a system that rewards the broker that produces the lowest premium. There are no points for quality, longevity, service, protection, prevention, or foresight. We've taken a stand against this industry norm because it's long overdue for a radical change and we don't believe it helps our clients.

Are you ready for this one? In reality, the insurance broker has very little to do with the size of the premium paid by their clients.

There are two much more important causes for a premium to be high or low. One is the insurance company and how competitive they want to be at any given time. Their competitiveness can include everything from having a slow quarter at the regional office to an underwriter needing increased production to make their budget for the year. Maybe a manager will get a bonus for top line growth (ignoring profitability), so he or she is able to lowball the premium. The second big determinant is simply the claim experience of the contractor. It has very, very little to do with any talent on the part of the broker.

The typical insurance agent or broker focuses on the price of insurance and keeping the customer happy in the short term, **not** the preservation of the customer's business. He or she doesn't bring any tough love to the relationship. Instead, they tell you, the client, what you want to hear (that they can get you the cheapest insurance), instead of what you need to hear. Meanwhile, you're thinking, "Phew, I have my insurance, everything's fine now." But, sometimes businesses really need to change what they're doing in order to protect themselves. If an insurance agent just focuses on keeping you happy by selling you cheap insurance and leaving you to think that you're fully protected, I think they're doing you a disservice. That's just a blind adherence to an industry norm that needs to change.

So, we ask the question philosophically, "Why should the broker be rewarded for something he or she didn't do?" What I say to people, somewhat tongue-in-cheek, is, "Whoever gets the stupidest underwriter wins." If you're reading this book, you understand the game where insurance carriers are allocated among competing brokers and, at the end of the day, the lowest price wins. What I'm going to suggest and reinforce throughout this book is that you have to look at insurance differently than you ever have before; you have to ask this question first, "Which combination of Risk Advisor and insurance carrier protect my blind side best?" Once that question has been answered, then look at the cost. And then look at cost even more broadly as I'll explain later.

Keeping this in mind, I think that most insurance brokers consider their jobs to be more like those of salespeople than like Risk Advisors. Their goal is to keep your business. In other words, because they get paid a commission they don't want to lose your business and lose their income, so they'll do whatever it takes, competitively, to keep your business.

I want to flip that concept on its head and say that my goal as a Risk Advisor isn't to keep your business; it's to keep you *in* business. If your business is multigenerational, then insurance protection is akin to family protection. You want to protect your company at all costs and you need a trusted advisor whose expertise is to help you do that.

Protect your blind side. Remember, you can't project your past experience into the future. The battle that I fight most often is with a contractor who looks at everything that's happened in the past and says, "Well, it's never happened before, so I don't believe it's ever going to happen." That's misguided thinking.

I have a friend who owns a construction company and also happens to be a ski patroller, and every year, I go to Vermont to ski with him. Now, I've been skiing for forty years and I've never been hurt. But in the chairlift my friend will point out, for example, a tree on a trail and tell me how someone hit it last weekend and how injured he or she was and what hospital they went to. Or he'll point out another spot and tell me how someone went flying, landed on someone else and put them in the hospital. I let him tell me a couple of his horror stories, and then I tell him, "Please, I don't want to hear any more." Because the wrecks, as he calls them, are so spread out among the twenty, thirty, forty or even hundreds of ski trails, most people never see them. But he sees them because he's involved in them as a patroller or hears the stories from his peers. It's easier, and much more pleasant, for me to ski down a trail without thinking that catastrophe is just around the next turn, but in reality, it is.

In my role as a Risk Advisor, I'm a lot like that ski patroller. I see the bigger picture, and even though you might not want to hear it, my role is to point out the things that I know happen to other people and to alert you to the probabilities. Catastrophes happen every day, so you need someone to help consider the consequences and prepare for them, even if it's a bit unpleasant and even if, especially if, it's a complete reversal from the way you've handled risk and insurance in the past.

2 Insurance: Just the Tip of the Iceberg

"I've never been in any accident of any sort worth speaking about. I have seen but one vessel in distress in all my years at sea. I never saw a wreck and never have been wrecked, nor was I ever in any predicament that threatened to end in disaster of any sort."

—E.J. Smith, 1907, Captain, *RMS* Titanic

It's interesting that I'm writing this book at a time where we have just experienced the most cataclysmic financial collapse in our generation, if not in history. I was watching the financial news yesterday afternoon, October 10th, 2008, and throughout the course of the eight-hour trading period, the Dow Jones Industrial Average covered a range of 1,300 points. Just like our Captain Smith of the RMS *Titanic*, no one ever would have predicted such a day on Wall Street. From a personal perspective, I hope it never happens again. But from a professional perspective, it's the kind of thing that a business owner needs to be conscious of.

A book was written in 2007 called *The Black Swan*. It's a book that I'm not sure I'm going to be able to finish because it's so technical and complicated from a probability standpoint, but it's something that speaks to the heart of this book and to what a Risk Advisor needs to be aware of when advising clients. The central idea of the *Black Swan* theory is that at one time, when the world was smaller, hundreds of years ago, nobody had ever seen anything but a white swan; and they thought that because they'd never seen anything else, that was all there was. But one day someone saw a black swan and realized that no matter how many times you've seen a white swan in the past, it doesn't mean you won't see a black swan in the future. Same as our friendly captain of the *Titanic*: he'd never seen a wreck, never been involved in one, didn't expect to be in one, and he was captain of the sturdiest ship to ever sail the seas. Then out of the blue, he hit the iceberg, and we have the most catastrophic sea disaster in world history.

What this all boils down to in terms of risk, and in regards to your business, is blindness with respect to random catastrophic events. When people look at insurance and risk management, they tend to see the pennies instead of the dollars. They're looking at the small, predictable events, and they're blind to that large, catastrophic event. I'll refer back to the twenty-one-million-dollar jury verdict for the cell phone/car crash incident. No one looking forward could ever predict that even the largest company could experience a claim of that magnitude for an incident of that type, in which a seventy-eight-year-old woman – who wasn't killed in the accident – would be awarded multimillions by a jury.

There are also *positive* black swan events, such as the advent of the personal computer. I think it was the president of Digital Equipment Corporation, Ken Olsen, at the time the PC was invented who said, "Who would ever need a PC?" And shortly thereafter, Bill Gates was saying, "I want to put a PC on every desk, in every office and in every home." That's something that never would have been anticipated back in the seventies, but now we can see in retrospect how normal, and brilliant, that was. Now, we can all say, "Of course!" And wonder how we ever lived without computers.

INSURANCE: JUST THE TIP OF THE ICEBERG

The negative black swan events can happen very, very quickly. It's much easier and much faster to destroy than to build, as we saw on Wall Street in October of 2008, when literally trillions of dollars of global wealth evaporated. A month ago, two months ago, a year ago, not even Warren Buffett, the financial genius of our time, would have predicted that this would happen in this way.

So I'm asking you to look at risk through a new lens. Look at the world that we're in today and think of it in the context of the *Titanic*. Are you sitting at the helm of your business looking at all of your past experiences and saying, "My past experience has been good, so I'm going to rest assured that it's going to continue into the future." That's a false sense of security, and you need someone who has the wisdom of experience, who like my friend on the ski patrol, has seen all the wrecks and can point them out to you so that you can be prepared. If that Risk Advisor is really good, he should be able to look around and help you put systems and practices into place that will prevent the wrecks altogether.

Total Cost of Risk

Total Cost of Risk (TCOR) is something that progressive insurance brokers are speaking about more and more with their clients. Like a lot of things in insurance and risk management, this concept began with the largest of companies, and a risk management consultant called Tillinghast came up with TCOR to help large companies benchmark their costs relative to risk. Where medium-sized contractors look at their insurance premiums or their insurance budget, the larger companies look at a bigger number that reflects all of the costs associated with the risks in their business; by definition this means insurance premiums, claims costs, the cost associated with loss prevention, and the cost associated with the administration of insurance and risk management programs. It is also viewed as the *Direct, Indirect, and Preventative* costs associated with risk. (See chart below, which is a copyrighted risk management tool by Sitkins International.)

Preventive	Direct	Indirect
Safety and risk management	Transportation expense	Employee turnover
Pre-employee screening	Production decrease and downtime	Reputation with insurance companies
Preventative process and procedures	Product or equipment loss	Vendor/supplier turnover
Website analysis and review	Replacement of first aid supplies	Senior management
Safety systems	OSHA fines	Loss of market share/ market shift
Safety reengineering	Consultant fees	Business failure of vendors or customers
Culture evaluation and management	Blood born pathogen issues	Warranty losses
Insurance premiums	Employee theft	Employee watching and discussing accident or events
Wellness programs	Market shift	Morale
GPS asset tracking and monitoring	Self insurance deductibles	Loss of reputation
Disaster recovery plans and systems	Self insurance gaps in coverage	
New hire orientation	Additional productivity loss of other workers	
Safety manager salary, expenses, taxes, etc.	Legal expenses	
Personal protective equipment	Loss of productivity after accident	
Safety assistant salaries, expenses, taxes, etc.	Management time to administrate injury activity	
General safety meetings	Other employees time to administrate injury activity	

Preventive	Direct	Indirect
First aid supplies	Active claims process	
Facility capital improvements for safety	Legislative updates/changes	
Maintenance and repair of safety equipment	Under value building/property value	
OSHA preventive compliance costs	Certificate of insurance management	
Safety committee time	Injured time	
Supervisor prep time for training	Medical visits while on light duty	
Annual reviews of TCOR systems	Decreased productivity while on light duty	
Training	Supervisor rescheduling and training	
Loss prevention and loss measure	Office preparation of reports and managing the loss	
Legal expenses	Reactive safety activities	
Vendor service plan	Overtime required	
Technology updates	Medical paid by employer	
Continued education	Clean up costs	
Cost of time on toolbox talks	Prescription costs	
Cost of materials for toolbox talks	Accounts receivables	

For the largest companies, those numbers are all added up and a ratio is created relative to their sales. You look at insurance premiums, claims paid and reserved, loss prevention expenses and administration expenses in relation to sales dollar amounts, and you come up with a ratio. The largest companies benchmark themselves by industry with those ratios. But for you, there are no statistics out there to help you benchmark this. However, you certainly can take a broader perspective on risk management in your company. Once you do, you'll start to manage risk differently. Once you understand the TCOR, again the Total Cost of Risk, as opposed to just the insurance premiums, that's when you should have a light bulb going off, your "Aha!" moment. All of a sudden you'll see that there are a lot of profit dollars leaking out of your business every day that you're not measuring or paying attention to, and that you could really improve the financial condition of your company by proactively understanding and controlling your TCOR.

Are you starting to see that risk management and its financial implications is much more than just insurance? One of the things I've tried to do in my career is look at the largest companies and see how they manage risk, and then, understand how many of their concepts or systems or behaviors are transferable to the middle market companies that I advise. At most middle market companies, risk management is either an unknown or a foreign concept. It's really about insurance management, and once again, that is just blind adherence to industry norms. Insurance people selling insurance have focused attention on insurance, so it's become much more about insurance management. Think about it like this: I think insurance management belongs in the purchasing or the controller's office, whereas risk management for a medium-sized contractor should be on the owner or the CEO's desk. The owner or CEO of a middle market contractor really needs to have his or her head wrapped around the concept of Total Cost of Risk and they need to be involved.

As I mentioned earlier, risk is broader than what insurance covers. The Total Cost of Risk includes both direct and indirect expenses. (See chart). Most construction company owners focus on the direct costs and the most obvious one there is the insurance premium itself; but you can also look at deductibles, legal expenses, medical visits that an employee takes while on duty, OSHA fines, consultant fees, etc. These are all examples of direct costs.

The indirect costs are more difficult to quantify. OSHA and other industry consultants have put a range on the indirect cost of an injury or claim at anywhere from a 1:1 ratio *to as high as a 20:1 ratio on the indirect cost to direct cost.* In other words, if you have a worker's comp injury that's quantified at five thousand dollars, that is not the Total Cost of Risk to your company. If the person who was injured is one of your most experienced foremen leading a team on your most important job or a job that you're behind on, and his skills and experience are essential, his injury will cost your company a lot more than five thousand dollars.

Another example is a larger, more catastrophic claim, maybe a severe fall, an electrocution, a partial amputation witnessed by many of your employees, which obviously will frighten them, distract them, and provide a lot of stress in your work environment in the succeeding weeks. There's going to be talk about that injured employee, his family, and the impact on finances. You may need to bring in outside support to help the crew. You'll have to hire and train a replacement worker.

There are many indirect costs that business owners are not conscious of, and a good Risk Advisor is one that will provide you with some kind of measurement tool to understand what those indirect costs are. Once you understand the indirect costs and how they work, you're going to build much more prevention behavior into your culture than you had before. Then, to that end, you're going to have increasingly more success, as compared to your competitors, as you eventually eliminate those indirect expenses.

Still, there's a lot of money left on the table, and the amount depends on your individual business and your profit margin. It depends on how many other severe claims you've had in recent history. But to emphasize the extreme, the hidden costs could be as much as twenty times the actual cost that you see on a loss run. When you look at your experience mod rate, or you look at the claim runs provided to you by your insurance company or your insurance broker, you can be led into a false sense of security that you're looking at *all* of the dollars when, in fact, you might just be looking at the tip of the iceberg.

What lies beneath the surface are the dollars drained from your company that weren't quantified before. Fortunately, this concept of Total Cost of Risk is becoming more widely recognized and more widely understood in the middle market business community. As it does, smart Risk Advisors and consultants are coming up with ways to quantify real amounts for individual businesses. Once you're able to quantify it, and once you're able to see these sources of leakage, you'll be able to design and build the prevention strategies that will prevent this leakage or minimize this leakage in the future.

In summary, the tip of the iceberg in the context of risk management has two elements: First, there's the black swan event that you can't anticipate, never expect, and possibly won't be able to pay for. Second are the day-to-day costs of poor risk management or claims that may seem inconsequential on the surface but have significant hidden costs that are draining profit from your company on a regular and invisible basis.

3 Safe, Healthy, Happy and Profitable

"While you make a ten-minute safety speech, two people will be killed and about 498 will suffer a disabling injury."
SOURCE: NATIONAL SAFETY COUNCIL INJURY FACTS, 2008

"On average, four workers are killed every day on US construction sites."
MARK AYERS, PRESIDENT OF THE AFL-CIO
BUILDING AND CONSTRUCTION TRADE DEPARTMENT

One of the most alarming things I see in my role as a Risk Advisor, is that, as a cultural imperative, safety isn't viewed as an equal to sales, quality, or profitability—or some of the more common cultural values of companies. Safety is difficult to implement. It's not understood, and it's often delegated to safety professionals as opposed to being an overall cultural value, or total company priority, driven by senior management. Once again, when you look at the Total Cost of Risk and you start to understand the indirect costs of an injury, you really start to understand why safety is so important. It isn't just important from a financial perspective; it's also important from a human perspective. After all, a business is composed of human beings and those highly skilled workers are the most valuable asset a construction company has.

Looking back at construction in 2008 and then forward to 2009, we can anticipate a difficult economy. We're already seeing layoffs, credit is tight, building projects are either being suspended or postponed and construction workers are going to be laid off. They are also getting older and becoming more prone to injury. Fewer and fewer young people are going into the construction trades, and this all adds up to a labor shortage that will be felt by every contractor.

If you are a safe place to work, if safety is a strong corporate value, then you are going to attract smarter, better workers who want to have long careers. Safety certainly has a huge impact on your finances. In an extreme case, the 20:1 ratio of indirect costs to direct costs that I mentioned in the last chapter translates into this: every dollar of claim could actually be *twenty* dollars of total leakage in profit from your company. Compute that into the operating margin of your construction company and figure out how many jobs you'd have to win and execute profitably to cover one five-thousand-dollar worker's comp claim.

I believe that the executive leadership team of a construction company must set the example of safety. That doesn't mean that you're going out and demonstrating safety on the worksite, but what it does mean is this: through your actions, through your behaviors, through your emphasis and through the actions, behaviors, and emphasis of your key personnel, everyone in your company understands that safety is a priority. This means that members of your executive team talk to the workforce about safety. They participate on the safety committee. They provide and invest in the best safety training and the best personal protective equipment. They are conscious of OSHA regulations. They are tuned in to safety, just as they're tuned in to quality control and timely execution.

There's always been a conflict in the discussion of safety between how much of it is situational and how much of it is behavioral. My own feeling in this argument is that it's almost all behavioral. Almost all accidents can be prevented or avoided. There are insurance companies and consultants who promote a concept of a zero-accident culture. Just like my friend in the ski lift, I have a bigger perspective on what happens out there, and I think medium-sized construction companies would be absolutely amazed to hear some of the statistics of larger, safer construction companies who operate for millions of man hours without a lost-time accident. Ultimately, education, training, and a culture of belief that all accidents can be prevented and avoided is probably the best, and the most important first step in attacking these indirect costs.

There's another "human" aspect of your company related to safety: employee wellness. You may find it unusual for an advisor like me to be talking to a construction company owner, like yourself, about the wellness of his or her employees – that's because it is. Wellness is a relatively new concept, but more and more studies are proving that there's a very strong connection between an employee's wellness and how prone he or she might be to injury. In an extreme example, it's probably easy to see how a construction worker who is young and fit and works out with regularity is not going to be as prone to soft-tissue injuries and lifting injuries as an older worker who's not as fit, is overweight and unhealthy, and in compensating for lack of fitness, doesn't use the proper procedure or methodology and gets hurt.

You can also look at wellness from the standpoint of someone who's under a lot of stress and is distracted. There's a concept in the medical literature called 'presenteeism.' Presenteeism is the opposite of absenteeism, or being absent. Presenteeism means employees are present in body but not in mind. It means they're distracted, and this is probably going to be a very common occurrence over the next six to twelve to twenty-four months. If you own a construction company, all of your workers are going to be concerned about being laid off. They're all going to be concerned about whether their retirement plans are going to be adequate. If they're late in their career, they're probably even wondering whether they're going to be able to retire on time because their retirement assets have just been cut by 30 or 40 or 50 percent. (By the way, this means that you're going to have older workers on your staff in the future.) Any of these reasons can contribute to employees being depressed or highly stressed. High stress can lead to distraction, fatigue and illness, all which make employees more prone to accidental injury in the workplace.

If you are the construction company that can minimize the impact of these things I've just described, if you can encourage wellness among your employees, if you can introduce them to concepts of wellness as more progressive construction companies are doing today, they're going to be less prone to work-related accidents. If you can provide them with support for mental wellness and stress-reduction such as budget and retirement workshops, for example, you will experience less work-related accidents. There is no question that health and safety are tied together.

Statistics show that the overall number of work-related injuries has been decreasing fairly steadily over the last twenty to thirty years. Some of this is related to OSHA initiatives, and some of it is related to the construction industry becoming more sophisticated about the financial and human costs related to injuries. Both sides have tried to put in preventive measures to limit those costs. However, we still see extreme levels of medical cost inflation. All of us experience medical cost inflation each year as we look at what it costs to buy our own individual health insurance. Compared to the cost of inflation, which has been relatively mild over the last five to ten years in the low single digits, medical cost inflation has often been in the high single digits and even into the teens. With the current economic climate, there doesn't seem to be any end in sight. Managed care, on the health insurance side, has done a fairly good job of keeping this down, but we haven't seen equivalent managed care initiatives on the worker's comp or work-related injury side.

> *"According to a recent study by the National Council on Compensation Insurance, the factor driving medical severity is the growth in the number and mix of medical treatments. The study compared 2001-2002 to 1996-1997 and found that* **the number of treatments for all diagnosis increased 45 percent**, *while the increase for injuries such as knee and leg sprains was as high as 80 percent. Through negotiations doctors have been cajoled into lower fees, and to counteract these lower fees medical treatments have increased. Think of it like squeezing a balloon. When you pinch one end, it expands on the other. Doctors are, for the most part, not following evidence-based treatment protocol as set forth by the American College of Occupational and Environmental Medicine, and by not doing so, we are witnessing medical treatment plans that are misdirected and unnecessary, coupled with increased levels of treatment, additional testing, and the high cost of doctors to run these tests." –Frank Pennachio, cofounder and Director of Learning at the* **Institute of Work Comp Professionals**

What this means to you as a contractor is that you have to work with medical professionals who follow **Evidence-Based Medicine** (I imagine that at this point you've never even heard of this concept much less know whether the occupational medicine clinics treating your injured workers practice it) and who understand what it means to minimize treatment and keep your workers healthy. As I said earlier, what I've tried to do in my position as a Risk Advisor is look at larger companies and look at the direction that they're taking, to study their new initiatives and see how much of that is transferable to the medium-sized construction company.

Earlier in my career, I read a story about DuPont—a household name to most of us. DuPont got their start back in the Civil War making gunpowder. As you might imagine, if you're going to be in the gunpowder business and you don't have an extreme focus on safety, you're not going to have a long and storied career as a company. We can all be grateful for the inventions that DuPont has brought us, and one of them is that corporate safety was a strong cultural imperative from the start.

In fact, DuPont has a safety consulting division in their company, and they're probably as progressive as any company ever has been in creating not only a safe culture for themselves, but also for the world around them as they share their ideas and practices with other large companies. DuPont's job sites and factories, which are tremendous, complex facilities with the potential for lots of accidental injuries, literally go millions and millions of hours with no lost workday accidents.

More and more, the largest contractors have also tuned into safety in a big way. I believe they do it because they understand the human element, but they also understand the profit element. The best example I know of is Turner Construction. Turner has actually turned safety, insurance, and risk management into a profit center, and they've done this through the use of contractor-controlled insurance programs (CCIPs) that control safety on the entire jobsite. In the end, this company reaps the rewards of a safe worksite. As you'll learn later in the book, these same concepts can be applied to middle market construction companies, and you can enjoy the same benefits that Turner does.

The High Return of Your Safety Investment

There are many different ways to look at why making an investment in safety can have high returns for your company. Probably the easiest way to look at it is through worker's compensation and the Experience Mod Rate, or EMR. From the safest to the least safe, contractors can have an extensively wide range in EMR. In my practice today for medium-sized contractors, I have seen companies whose EMR is as low as 0.6, in the Connecticut marketplace, to those that are over 2.0. Anyone down at a 0.6 or 0.7 level has a huge competitive advantage in the marketplace. They are saving hundreds of thousands of dollars and potentially millions of dollars a year.

Larger owners of projects and larger construction companies are getting more attuned to safety all the time. For years, many of them have a stated rule that any contractor with an EMR over 1.0 will not work on their job, which is also probably the *threshold of fear* for most contractors. If you look at it from that perspective, the inability to bid a job, you can then look at your profit margin and your average job size and calculate the lost profits or opportunity cost of having a poor safety record.

Let's say for example your average job size is five million and your profit margin is five percent. Five percent is $250,000, so for every job that you're unable to bid, that $250,000 goes out the window. If you've got five of those, in which the owner or general contractor has that 1.0 EMR threshold and you can't bid, then multiply $250,000 times five, and all of a sudden it's not just the fact that your EMR at 1.10 versus your EMR at 0.82 costs you $65,000 in insurance premium. All of a sudden, it's ***$1,250,000 of lost profit*** because of the GCs or owners who wouldn't accept your bid.

When you think of what you could do with that cash—investing in your business and investing in your future—all of a sudden you get a handle on what this Total Cost of Risk means and how safety can become very profitable. If you had invested $50,000 or even $100,000, knowing that it would return you over one million dollars in profits on the jobs you executed successfully during the next 12 months, you'd do it in a second. It's the job of your Risk Advisor to bring you the tough love that explains this, calculates it so you really understand it and helps you make the proper investment. Making safety a strong cultural imperative in your company can happen when that $100,000 safety investment is revealed as a necessary stepping-stone to one million dollars in profits.

4 Insurance Gambling: The Risk Transfer

There are a wide variety of items and events that can have an impact on your business that aren't going to be covered by insurance. We have to start with how broadly you define risk and what kind of advisor can help you with that broader definition. Typically, most medium-size contractors define risk through the prism of their insurance policies. And it makes sense. Because this whole concept of risk is always discussed with insurance people it's always in an insurance context. But, larger companies have expanded the context of risk far beyond just what's covered by insurance. Risk is much more than the sum of your insurance policies.

Let's talk about insurance risk transfer. When you lose a profitable job due to a high EMR it's because you haven't been operating safely in the past, it's because the people upstream from you – a general contractor, or if you are a general contractor, an owner – look at you and say, "We think you're going to be unsafe in the future, and we won't hire you because you're going to contribute to our costs." Another common example is when contractors sign a contract with language to the effect of, *"any and all risks associated with."* Whenever you see that 'any and all,' it's a red flag; you need some interpretation, qualification, or limitation. **An "any and all" could be one very small clause that leads you right to bankruptcy court, because in many cases you are assuming 'any and all' risks not covered by insurance.**

There was a time in the not so distant past when contracts were simpler, insurance was simpler, and you could rely on standard language and terminology to protect your balance sheet. What I like to say now, in regards to this area of Contractual Risk Transfer (CRT) is that it's a game of musical chairs. Essentially, CRT means that one party is transferring risk to another party by means of a contract. This happens between owner-developers and general contractors, it happens between general contractors, construction managers and their subcontractors, and it happens between subcontractors and their sub-subcontractors. The typical areas of CRT are indemnification and hold harmless clauses, waiver of subrogation requirements, additional insureds, Owners and Contractors Protective Policies, Railroad Protective Policies, and other requirements to purchase additional coverage or modify language to existing policies. This game of musical chairs is all about who is going to end up holding the bag, which party will be exposed to the biggest risks when something goes wrong.

When you look at the financial impact of claims from construction work you're looking at big dollars, millions of dollars, tens of millions of dollars—and every party to the transaction wants the other party to pay the bill when something goes wrong. Obviously, some parties to the transaction have less leverage than others. And this is where the insurance companies enter the game.

The Risk Advisor's role in this is to interpret the risk transfer documents and guide the client to make the best choices; and once the client has made those choices to design insurance protection in such a way that any liability assumed can be transferred to an insurance company. Yes, there are some risks that cannot be transferred and there are some risks for which insurance companies will not provide coverage; but at least with an intelligent Risk Advisor, a contractor will be making these choices with his or her eyes open. He or she can balance those choices with the profitability on the job and the probability of getting stung with a claim they're going to have to pay for out of their own pocket.

In this game of musical chairs, one of the more significant areas that medium-sized contractors are not fully aware of is how much insurance companies have restricted coverage by introducing new language in their coverage forms. What I try to explain to my clients is that insurance carriers aren't doing this because they are trying to put their clients out of business; they're not doing this because they're bad people or for any reason other than that they're running a business. And just like a contractor running his or her business, insurance companies are trying to limit their liability. Even though insurance companies are in the business of accepting risk from their policyholders, they don't accept the transfer of anything and everything.

The unfortunate situation is when the contractor doesn't understand the intent of the insurance company – thinking the game is still being played by the same rules and that the coverage offered to them yesterday is the same offered to them today. This is clearly not the case, and even the largest construction companies have to fight for what they deserve in terms of coverage. They, too, need a Risk Advisor who has the perspective and understanding to know what to ask for, when to ask for it, and how much should be charged for it. If you don't have a Risk Advisor who understands these issues, you're going to lose that game of musical chairs. You will be the last one standing, and that could lead to the loss of your greatest asset, your business.

The simplest example I can use to illustrate is how *additional insured* language has changed in recent years. There was a time when almost all insurance companies used the same few endorsements to modify coverage to add additional insureds to liability policies. Today, in 2009, there are probably 40-50 different endorsements, some industry standards, and others proprietary to each individual insurance carrier. When you add in the non-standard insurance companies and the umbrella liability carriers, you might have double that amount.

A typical scenario on a construction contract is where there is a requirement that **_completed operations_** coverage be provided to an additional insured for 3-5 years post project completion. It is very common for an additional insured endorsement to provide **NO** completed operations coverage to an additional insured. If your insurance is written this way, you've got a problem.

Here is the strongest statement I can make regarding this issue and the danger to your business. My company has examined the insurance programs of over one hundred and fifty construction companies in the last five years. We haven't yet found one whose insurance program provided the right coverage to comply with all their contracts. In many cases, we have discovered that they are in non-compliance on every single job they are on. That goes to show how widespread the ignorance is on this topic because these contractors are providing Certificates of Insurance to some of the largest contractors in the country and no one realizes that the emperor has no clothes on.

Large companies have begun implementing the executive role of Chief Risk Officer (CRO); by definition that person is responsible for identifying, analyzing, quantifying, and determining the treatment for the risks that a company faces. And, yes, you read that right; the best and brightest construction corporations have added the CRO position to their executive team – right alongside the COO, CFO, and CEO. A medium-sized construction company has no one on their staff qualified to sit in this position. Most insurance agents and brokers are not qualified to act as Chief Risk Officer, and even if they were, what methodology or system would we have to determine their qualifications? If you don't have a Chief Risk Officer and you can't train one, who is going to perform that role? How are you going to determine if the agent or broker or Risk Advisor has the qualifications to assume that role?

Going All In

When you're playing poker and you go "all in" it means that you put all of your money in the pot. If you win, you win big. But if you lose, you lose it all – game over.

In the world of risk management, going all in means you put your insurance out to bid, you spreadsheet the insurance carriers and you pick the broker that brings you the insurance carrier who sells you the cheapest insurance. You have gone all in with your bet that that combination of broker and insurance carrier is the one who is going to protect your blind side properly. While this decision may seem to save you money up front, it's actually a very risky idea because if you've made your choice strictly on the price of the insurance, you have no idea what the qualifications are of the broker or the insurance carrier, their staff, or the quality of the coverage that they've put in place to protect your blind side.

I see this risky, "all in" behavior literally every time we look at a new client opportunity. One example that always comes to mind involves a construction company that is now a client of ours. When we first analyzed their operations, they had fifteen different legal entities, which is not uncommon for a contractor, a contractor-developer, or a GC subcontractor. This company was a kind of combination of the three. What we discovered is that five of those legal entities were covered by the primary liability policy, but not the excess liability policy; another five were covered by the excess liability, but not by the primary liability; and, five weren't covered by either one. To be clear on this, five of those legal entities only had excess liability coverage, which means they all had a million-dollar deductible; five of them only had the pri-mary million-dollar coverage so they had no excess coverage for a cata-strophic loss; and five of those legal entities had no liability coverage whatsoever. Hands down, the worst case I've ever seen of a company that went all in with a broker who was their friend but who was com-pletely unqualified to protect their blind side.

So that was the worst example, but there are countless others. Two contractors I know started new jobs that brought them into the arena of railroads, which, without getting too technical, requires spe-cific actions relative to liability insurance. Contractors who work near railroads must always accept a transfer of risk from upstream parties relative to that railroad work, and they need to modify their insur-ance accordingly. These were both very large and sophisticated middle market contractors, but they had no coverage relative to the risk they'd accepted on a railroad job, which means that minimally they could have been thrown off every single job they were on, but more impor-tantly that they were exposed to an uninsured loss which would have been catastrophic to their company. All because they went all in with an insurance company that sold them coverage but had no idea how to protect their blind side.

The Paradox of Insurance Buying

I mentioned earlier about this focus on bidding and comparing insur-ance programs based on the premium. The paradox of insurance buy-ing is that the more a contractor focuses on the price of insurance, the less protection they will ultimately have.

There are a few different reasons for this. If you look at the ex-ample I just gave of the construction company with fifteen different legal entities, each one of them covered differently than the next, my bet is that each time their company grew and they went out to bid, they only looked at the price of insurance—and they had an agent who didn't counsel them any differently. Their errors were compounded, these new entities were forgotten or overlooked or misunderstood, and all the focus was on the price; the focus on protection and risk management went out the window.

As you know, every company only has so much time, money, and resources to allocate to any task or project. If you allocate your insurance time or your risk management time to the insurance buying process, to sitting down and meeting agents, meeting insurance carriers and going through that whole bidding process, that means there's that much less time you have to devote to safety, analyzing indirect costs, proper medical treatment and return to work programs for your injured employees. Chances are, if you think all the gold is in the bidding process, you're probably completely ignoring all the other risk management issues to your detriment

The paradox of insurance buying is that the more you focus on insurance buying and obtaining cheap insurance the more expensive that insurance will become right along with your TCOR. Because you aren't focusing on preventing losses and managing the losses you have and eliminating the possibility of catastrophic loss, all of a sudden your loss ratio is out of control and so is the cost of your insurance. Insurance bidding is fool's gold because construction company owners will bring in a group of insurance agents, competitively bid their insurance program, and go through a very convoluted exercise saying, "Wow, look at that. We saved fifteen percent or fifty thousand dollars. It was worth it. We should do this every year." But just because you've saved money on your insurance doesn't mean you've saved money in the long haul. As I'll emphasize over and over again, you have to look at the Total Cost of Risk. Looking purely at insurance premiums tells you nothing about indirect costs, it tells you nothing about lost profits, it tells you nothing about the big dollars. It's really a focus on the pennies instead of the dollars. It's really a focus on today and not tomorrow.

If you make a decision to choose an insurance broker based on the luck they had in choosing the carrier that came in with the cheapest insurance, you very possibly could be getting the worst broker in terms of the advice, counsel, and risk-reducing services that they'll bring you in the risk management process. It's very common to have an insurance carrier who may be the right one and provide all the coverage and unique services you require in combination with a broker who is poorly matched for what you need. It's extremely important that you don't end up with what I call the **Right Carrier/Wrong Broker** scenario. You have to have the right Risk Advisor or the right broker *and* the right insurance carrier. You have to have both, no exceptions.

I mentioned earlier that insurance management belongs in the purchasing or controller's office and that risk management should be on the owner's or CEO's desk. I think it is critical for construction company owners to understand this next concept.

Insurance management is always tied to insurance policies, and insurance policies in terms of purchasing decisions are tied to their expiration dates. As every reader knows, every insurance policy runs for a twelve-month period and has an expiration date. You, as a company owner, tie your insurance decisions to this arbitrary date, when, in reality, this is an executive issue that should be examined any time during the year and handled appropriately.

Put it this way, if you discovered that one of your trusted advisors – your lawyer, your CPA, your banker – was incompetent (and what I mean by incompetent is that they're doing things that could cost your company money or maybe even cost you your business), would you wait until some arbitrary date in the future to replace them or would you deal with it immediately?

If you can't predict the future how do you know that today is not going to be the day when disaster strikes? If you discover that your Risk Advisor or insurance agent is incompetent today, do you think it makes sense to wait two months, three months, four months, eight months, until whatever your normal insurance anniversary date is to hire a competent advisor? My advice to you is to find someone who will start working for you immediately. Hiring a Risk Advisor should never be tied to some arbitrary date. Maybe you discover it by reading this book, maybe you discover it when something goes wrong, but you should make an immediate change if you find out that whoever is handling your insurance has only gotten you the cheapest deal and done nothing about understanding and helping you manage the risks of your business. I am endlessly perplexed as to how we ended up with a system where otherwise smart business owners bet their companies on inexperienced or incompetent insurance salespeople peddling cheap insurance.

5 Not Every Claim Is Created Equal

If you can reach back to 1992, you'll remember that Bill Clinton unseated the first President Bush with the help of the term, "It's the economy, stupid," because it was always about the economy – whatever 'it' was. In the world of risk management and insurance, my rendition goes like this, "It's the claims, stupid."

What's lost on most risk management practitioners is that it's all about the claims – always has been. We've discussed the Experience Mod Rate (EMR) with worker's compensation, and anyone who has purchased worker's compensation understands that formula pretty well. It's a three-year running average that compares one company's claim experience to all the other companies who do the same thing within a state. What I think many insurance buyers don't realize is that every type of insurance you buy is experience rated. We all understand that this is how it works with car insurance, right? If a teenager gets into a car accident, the insurance premium will go up and there may be loss of driving privileges. But, whether you buy worker's comp or automobile or general liability or absolutely any form of insurance coverage, it's also going to be experience rated.

As I've said before, whoever has the lowest number of claims wins. If every policy you buy is experience rated, just like worker's comp is, then claim prevention should be the biggest priority for any company or any senior management team. That is the way to truly get the least expensive policy. No amount of insurance bidding can erase a poor claim record, so the benefit of having fewer claims is that you will pay less money for insurance whether the market is hard or soft. Regardless of economic circumstances, those who have the lowest claims will pay the least amount of money for insurance. If you can eliminate a claim, that's going to have the most financial impact on your company.

The last point I want to make in terms of a claim history is that insurance underwriters talk about frequency claims versus severity claims. That's a relatively simple concept. A frequency claim is the automobile fender bender. It's the medical-only worker's comp claim. It's the slip and fall general liability claim. It's the claim that's not going to involve a lawsuit. It's probably going to be settled for under five thousand dollars, and it's the one that goes away quickly and easily.

However, there's a rule or an axiom in insurance that frequency leads to severity. The concept is that if you have lots of small claims, insurance companies realize that there are a lot of near misses, and those near misses could be small or large claims. So in the end, prevention should be the absolute highest priority of any construction company.

The severity claim is the one that's going to cost a lot of money. If you're a contractor who has very few frequency claims, but all of a sudden you have a severity claim, there's a chance that severity claim will be discounted – the underwriter may look at it and say, "Wow, that's really a one-off, highly unusual circumstance. It's not going to happen again, and I don't foresee that in the future." Other times the underwriter looks at a severity claim and sees danger – the severity claim serves as a big red flag. In this case, an underwriter needs a lot less imagination to look at the future and see what another severity claim might cost them.

Oftentimes, claim and claim management are relegated to the financial part of the organization, or the human resource part of the organization. Many times senior management is not aware of claims until they become really large numbers that have a major impact on premium. But, this is completely ass-backwards.

Senior management needs to get involved in claims at the earliest possible point in time. (When someone is injured at DuPont, which has 60,000 employees, the injury report is on the CEO's desk in 24 hours). Most importantly, as I said earlier, risk management needs to be on the owner/CEO's desk. The CEO needs to know when risk management goes wrong and understand that each and every claim is evidence of their risk management's failure. Secondly, you can't prevent a recurrence of a situation if you don't know that it happened in the first place. Senior management should be spotting trends. If they can spot a particular type of work, jobsite, or maybe even a particular foreman who's not enforcing safety at the highest level, they know where, who, and what to fix.

Most medium-sized construction companies leave this job of claim management, claim identification, and trend spotting up to the insurance company. Insurance companies have competent claim and loss prevention people on their staff. But I have to counsel that when an insurance company collects their premium dollar, there's only so much of it that they can allocate to risk prevention or risk mitigation. I wouldn't advise any construction company to rely solely on their insurance agent or their insurance company in this matter.

The typical insurance company and agent are more reactive than proactive. They're not going to determine cause and effect or help you build prevention into your operations. They're going to explain to you after the fact why you're paying so much more this year than last year. And, I'm guessing that's not what you want to hear. I'm also going to assume that within your own company you probably don't like surprises; you probably mandate that your senior managers bring issues to you before they blow up so that you can react to them early enough to minimize the impact. The same is true with insurance. *You need a Risk Advisor who is proactive, someone who sees these things before they happen and can prevent them before the damage occurs.*

Risk Mapping and Prioritization

Risk mapping is a concept employed widely by larger construction companies. Essentially, it involves looking at all the possible things that could happen and creating a risk map as a visual tool. Most construction company owners look at the insurance coverage they've always bought and think it is what they should always buy. I counsel my clients instead to look at the probabilities of what might happen, create some analogies, and then look at the kinds of risks that scare you the most.

I've been counseling construction companies for thirty years, and I can count on one hand the number of times a construction company has had a liability claim in excess of one million dollars. Yet almost every client I have has excess liability in the amount of five, ten, fifteen, even one hundred million, and I think it's smart that they do, even though the likelihood of needing it is very low. However, I've tried to get them to consider other types of insurance protection against situations I think can threaten their companies that are of a much higher probability. In the past, they hadn't considered them, so they were still buying very low-deductible property or automobile collision insurance, both of which they could easily absorb, instead of mapping out the types of risks that could destroy the company.

I come to my profession with the perspective that my job is to protect the blind side, to preserve at all costs the company and what has been built over those multiple generations. You can't do that without risk mapping. You can't do that without finding or coming up with some methodology or practice that allows you to allocate limited premium dollars to those things that have the highest probability and the most severity.

6 Make a New Plan

In Stephen Covey's famous book, *The Seven Habits of Highly Effective People,* one of his chapters is "Begin with the End in Mind." In other words, think of what you want to accomplish or what the end goal is, and build a plan to get there. When you consider your buying behavior with insurance, as I've said repeatedly, it's typically about paying the lowest premiums. You haven't looked at the Total Cost of Risk; you haven't formed a plan to improve that Total Cost of Risk.

I'm asking you to forget your past buying behavior and start by beginning with an end in mind. What do you want to accomplish over the next twelve months relative to risk management? What are you going to measure in terms of quantity and quality? How are you going to measure success or failure? What are the benchmarks going to be? How are you going to hold people accountable?

The answers to those questions are going to be different for every company. If you've been doing well obviously you're talking about managing small differences. If you've had some serious claim problems that threaten the cost structure of the company and maybe even threaten its future because you're paying so much more in terms of direct and indirect costs than you should, then you really need to take drastic action. You have to start out saying, "Where am I headed?" "What kind of a plan am I going to create?" A Risk Advisor is different from an insurance agent in this respect. An insurance agent usually provides insurance coverage and reactive service, a Risk Advisor sits down with you at the outset and creates a plan with you to make improvements. What differentiates an insurance agent from a Risk Advisor is that the Risk Advisor's primary role is to *Design and Implement Risk Reducing Plans* vs. the insurance agent who simply places and services insurance policies.

If a company has a claim frequency problem, simple prevention tactics are going to have a big impact. If a company has severity claim problems, the plan will be much more complicated and will involve an examination of the individual circumstances of those losses and whether or not they can be prevented in the future. A severity issue might have a much longer-term plan than a frequency issue. For example, if your EMR is 1.20, it could take several years to change the company culture and get the claims reduced to more reasonable levels.

I've come up with a concept that I think every Risk Advisor should adhere to: *we don't get paid unless improvement is made* (more on this when I discuss Performance-Based Risk Management). As I said earlier, I think that our industry is broken. I think that paying insurance agents to place insurance with insurance companies on a commission basis is the wrong way to compensate them, and it doesn't show any measure of their effectiveness – it certainly doesn't encourage them to protect their clients. Compensation should be performance based and it should have goals. It should resemble your plan for annual employee compensation and bonuses. If your employees aren't being productive or performing well in their company position, then they're not going to get recognized or rewarded.

Risk Advisors versus Insurance Agents

Our industry has held to the insurance bidding process for a long time, and you, as a construction company owner, might not understand just how expensive and labor intensive it is when an insurance agency engages in the bidding process. It takes a lot of time, it takes a lot of money, and it takes a lot of people; and, if an insurance agency has allocated that level of resources to the bidding process, you better believe they don't have much left over to allocate to proactive risk management.

My company is markedly different than most in that we flipped this model on its head about five years ago and decided to shift all our extra dollars, or all the dollars that we had in the bidding process, to providing risk reducing services to our clients, to focus our time, our money, and our resources on reducing the Total Cost of Risk instead of insurance bidding.

Claims determine the premium, whoever has the lowest claims wins, and ultimately, whoever hires the advisor that can help them achieve the lowest level of claims is making the smartest decision – a much smarter decision than choosing to go out to bid every year. Just think about the resources you have to allocate to this bidding process every year.

The world is getting broken down into those brokers or Risk Advisors who have changed the model and are helping their clients prevent and/or minimize claims and those that are just helping their clients buy cheap insurance. If you've got a low level of claims, the most inexperienced person on your insurance agency staff can get you cheap insurance, but it'll take the smartest, most energetic, most insightful, and most creative one to help you design and implement a program to reduce your TCOR.

While insurance companies are only going to allocate so much of every premium dollar to help you with your claims and prevention, my organization learned a long time ago to supplement those insurance carrier services so that our clients can achieve the best outcomes. We offer safety services, human resource consulting services, loss prevention services, online technology tools, legal advisors, assistance with safety meetings, wellness programs, pre-employment screening, and disaster recovery plans, among others. And, believe it or not, we offer them for the same amount of money that the commission based insurance salesman is earning, i.e. his commission from your insurance plan purchase.

Many times when we describe what we offer to a perspective client, they don't believe us. Our value proposition and business model sound too good to be true. They ask, "How can you offer so much and get compensated the same way as other brokers?" The answer is simple. We don't earn our new clients by quoting premiums. Ever. We have an ironclad rule. We only quote for clients, not prospective clients. You pick us to be your Risk Advisor not because we provide cheap insurance but because we will design and implement a plan to reduce your TCOR. Reducing TCOR for a client is a long and expensive proposition. We believe our competitors invest their time, energy, and resources in bidding on new business. We invest ours in reducing the TCOR of our clients. That is a black and white, fundamental difference between our Risk Advisor's approach and that of a traditional insurance agent or broker.

Most contractors have never designed a risk reduction plan. They wouldn't know where to begin. It's never been offered to them, and they don't understand the importance of it. That's where we come in, to help you work through the questions that need to be asked as you set out to design your company's plan.

Beginning with the end in mind, what do you hope to accomplish? Who should be on your team, the Risk Advisors and/or the insurance company? Where are you going to focus your attention? If you're a contractor, are you going to focus it on one jobsite or all the jobsites; do you have problems in one place that you don't have in another? When are you going to begin? Do you begin immediately because the issues require urgent attention? Why are you doing this? What's the financial impact? How are you going to quantify it? How are you going to implement the plan? How are you going to know if you're successful? The answers lie in the ability of your Risk Advisor to use a systematic approach to designing and implementing your plan.

At our firm, we use something called the Risk Reduction Approach™. My company belongs to Sitkins International, the most prestigious network of insurance brokers and Risk Advisors operating throughout North and South America. Under the guidance of consultant Roger Sitkins, this organization has developed a proprietary approach to reducing and managing TCOR. It's a unique process that allows us to build a plan and identify measurable goals that will have an impact on the TCOR for your company. A plan like this needs to be updated at least annually, if not more often, and responsibility for the execution of the plan is shared between the employees and management team at the client company along with the Risk Advisory firm engaged to assist with the plan.

If you look below, you'll see mistakes that I think are made by construction companies that can be addressed in a risk reduction plan. This certainly is not all-inclusive, but it demonstrates many of the common mistakes that I see in my practice, and could form the basis of how you plan your goals.

Mistakes Contractors Make Relative to Risk Management and Insurance

- Waiting until it hurts before taking action
- Cheap insurance as a goal instead of financial security and reducing Total Cost of Risk

- Not being proactive about safety

- Not being proactive about claims

- Ignoring subcontractor Certificate of Insurance issues

- Not understanding Contractual Risk Transfer (CRT)

- Exclusive focus on price of insurance
  instead of Total Cost of Risk

- Ignoring the experience/knowledge of the competing brokers

- Ignoring the depth/breadth of the risk reducing
  services offered by competing brokers

- Having a "that will never happen to us" attitude about risk

- Safety culture not supported by senior management

- Deficient hiring practices

- No tools/training in claim procedures

- Failing to establish relationships with clinics
  that will treat their injured workers

- Not being supportive of injured workers
  to get them back to work quickly

- Paying their broker large commissions and
  getting very little value in return

- Rewarding surety brokers by letting them handle risk
  management which they are unqualified to do

- Ignoring high risk exposures because they have
  never had that type of incident before

- Hiring a generalist insurance agent vs. a specialist Risk Advisor

Implementation

Certainly as a contractor you understand that when you build something there's always a timeline and a need for accountability to finish the job on time. There is no difference between that and a plan to reduce your Total Cost of Risk. Depending on what project is being worked on, there can be monthly milestones, quarterly milestones, or semi-annual milestones, but all along the way there has to be some type of measurement of success and accountability for making it happen.

The Risk Advisor needs to be the lead in implementation. He or she must set reasonable benchmarks and assist you, the client, in pushing these goals along throughout the period of engagement. However, Risk Advisors cannot execute a plan like this on their own. The primary responsibility is always going to be with your company because ultimately it's about prevention, it's about claims, it's about the things that happen in the field and on the jobsite – and those are the sole responsibility of your company.

7 Show Me the Money

Our definition of the Total Cost of Risk is Premiums + Claims + Prevention Costs + Administrative Costs. The most important criteria when picking a firm to advise you on risk should be, "Who has the resources to have the greatest impact on my Total Cost of Risk?"

The first criterion is, "Which firm has the capability of protecting my blind side?" All the focus on prevention and claim and premium minimization won't mean anything if there's a hole in your risk management that could potentially destroy your business. So ask if this firm is capable of protecting your blind side? Do they have the knowledge and wisdom to understand the risks that a construction company faces and how they can help you protect your company? Have them explain their answer if it's yes.

The second criterion is the focus on minimizing the Total Cost of Risk, with the primary emphasis on prevention and claim minimization. Ask them what kind of resources they bring to the table—what kind of people with what kind of experience and credentials, what kind of systems, what kind of tools to prevent and minimize claims?

Unfortunately, in our industry, good service has traditionally been equated to good *reactive* service. In other words, someone always returns your phone calls, someone always answers your requests on time, someone is always there when you need to ask a question about coverage, et cetera. My feeling is that those are minimum standards of acceptance to even play in the game of risk management. However, good reactive service is not indicative of the *proactive* service required to design and implement a plan to reduce TCOR. See "**Menu of Potential Project Work**" in the next chapter to compare the service capabilities of Risk Advisors vs. typical insurance agents/brokers.

Professionals or Amateurs? You Decide

Insurance agencies and brokers, by definition, are typically called "sales organizations." The reason for this is that, for the major insurance companies they represent, they are the primary source of distribution. As a result, there is a bias toward making new sales versus providing the best possible risk advice and proactive risk-reducing services.

If you look at the insurance distribution industry thirty years ago, most insurance agents or agencies were small, community businesses. With the exception of the large national/international brokers who focused on the multinational companies, the rest of the industry was small and community-based. Because of this, they knew their community, there was a high-touch focus with their clients, and they provided a high level of consultative advice.

As the industry has evolved it's become an industry of larger companies. There are even many insurance brokerage firms that are public companies. Many are part of roll-ups, and a number are now owned by private equity companies. These larger firms aren't as close to their customers. They have a much greater emphasis on top-line growth and return on investment or return on equity, because investors own them. There isn't the focus on the quality of the risk advice to the customer or the client, and some of these firms can have a very high turnover. They hire salespeople who can generate sales but aren't providing the level of advice that is required.

If you're a construction company focused on Total Cost of Risk and protecting your blind side, how do you evaluate an insurance firm differently than you have in the past? I believe that the first place you have to start is with that one individual who's going to be ultimately responsible for providing you advice. The comparison I draw is to look at a law firm, or a CPA firm, and the relationship that you have with them. If you have a serious legal matter involving complex contract litigation, estate planning, or anything of significant impact, financial or otherwise, to you or your company, you probably have a senior partner involved. The same is probably true with your CPA firm. Your financial statement is vital to your success as a contractor, vital to your success in securing surety bonds. You want nothing less than the highest level of advice from that CPA firm you've hired.

Why then do smart contractors choose insurance agents or brokers who have little or no experience, who are more amateur than they are professional, trainees instead of senior partners?

Do you think you're getting the same level of advice from a twenty-four-year-old who happens to arrive at your doorstep with the cheapest insurance policies? Do you think that person is the equivalent of the senior partner in your law firm or CPA firm? You are trusting this adviser to protect this business that you've built with your own sweat and blood, or your parents' or your grandparents', and now you've turned the protection over to somebody who is one or two years out of their training. They've never been challenged. They've never faced complicated problems. They have very little perspective on the worst things that can happen and how they can advise you. I have been a practitioner for over thirty years, and without sounding too self-serving, I can tell you that you need someone with the most amount of experience working for you if you want your blind side protected properly.

Let's return to the Lawrence Taylor analogy for a moment. Over the last twenty-five years the left tackle, who protects the blind side on a quarterback, has literally become one of the most valuable players on the team when you look at how they're compensated. On many football teams, the compensation of the left tackle is higher than almost every player except the quarterback. The reason for this is that they are protecting the most valuable asset on the team, which is the quarterback, in an offense that is primarily conducted as a passing game in the modern football strategy. If your business is your most valuable asset, you've got to have the most valuable person you can find protecting it. Not the cheapest, not the least experienced, not the one who has the lowest cost insurance, but the one who can do the best job of protecting that income-producing asset. And that is, in most cases, going to be the senior partner, the experienced Risk Advisor – not the guy who got you the lowest premium.

Beware of Surety Brokers Providing Insurance Advice

Another common mistake of contractors is relying on their surety professional for risk management advice. In many cases, the surety professional is the lead relationship between a construction company and their insurance brokerage firm. In many cases, the surety professional not only provides advice on bonds but also on insurance. My experience has been that this is one of the biggest mistakes a contractor can make. Insurance, risk management, and surety are all different areas of expertise. It's easy and necessary to combine insurance and risk management, but bonds and surety are a completely different animal. The biggest mistakes I've seen, and the biggest holes in insurance programs, are when surety professionals overstep their area of expertise into the realm of insurance. A surety professional has never been focused or trained in the area of risk management.

If you look at another profession, any other profession – I'll use law and accounting again – they're all specialists. If you go to a law firm, you certainly don't choose the same lawyer to do your estate plan as you do to defend you in a contract dispute. The same is true here. There is an area where surety professionals understand the issues of risk management, but they're not conversant in the nuances of loss prevention, insurance policy language, or claim mitigation, which are all key elements of Total Cost of Risk.

CAUTION:

With very few exceptions, your surety broker should **NEVER** be your Risk Advisor (yet this is a very common practice). There are some very notable insurance brokerage firms, local, regional and national, whose entire construction insurance practice is run by its surety professionals. This is absolutely nuts, especially for you, the client. It's like the tax CPAs doing the audits or the trial lawyers doing real estate closings. Insurance and surety are distinct and separate disciplines which should never be combined in their delivery. The insurance companies treat them as completely separate operations. There is absolutely no overlap. Why should insurance brokers be any different?

Here is a guaranty I will make to any reader. If your insurance program is handled by a surety professional, let me come in and do an analysis. If I don't find coverage and/or TCOR management gaps that you could drive a ten-wheeler through, I will treat you to the best steak dinner in town.

8 The Test Drive

In almost all cases, insurance decisions are made in an all-or-nothing context. As I explained earlier, once a year the insurance policy comes up for renewal, and the construction company owner decides if he or she is going to choose a new broker, new insurance company, or remain where they are. Typically, that decision is based on the price of the insurance. More and more, as construction company owners become more sophisticated, they're looking beyond just the price of the insurance; they're looking at the capabilities of the firm. However, they're still basing that decision around the timing of their insurance policies, and I think this is a mistake.

If you discover that your Risk Advisor or insurance agent is incompetent, you should change him or her immediately; you shouldn't wait until the anniversary date of your insurance. Minimally, you should hire a Risk Advisor to supplement the agent that you've found to be incompetent, so you've got a backup plan until that insurance can be changed.

A great way to evaluate a potential Risk Advisor is to ask him or her to do some project work before your annual anniversary date, so you get some sense of how qualified they are and what kinds of capabilities their firm brings to the table. If a company is just an insurance agent and primarily engaged in the service of providing insurance and insurance only, they wouldn't even know where to begin in terms of engaging in project work or doing consultation. If a company is truly a Risk Advisor, they'll have an entire menu of services (see examples below) that they can provide, and they'll offer any or all of them on a project basis so you can measure their effectiveness and measure the knowledge of the people that are engaged on that project for you. This is a much better way to gauge the qualifications of a prospective Risk Advisor firm than to wait until your anniversary date.

Menu of Potential Project Work to Evaluate Risk Advisors

- Sexual harassment training
- Evaluation of HR policy and procedures to comply with state and federal laws
- OSHA safety audit of construction sites
- Audit, update, and revision of safety manual
- Audit, update, and revision of employee manual
- Assessment and recommendation to improve return to work methods
- Comprehensive risk survey of all operations by trained risk management expert
- Compliance audit of all subcontractor certificates of insurance
- Rewrite insurance/indemnification sections of construction contract
- Seminar for project managers on Contractual Risk transfer
- Seminar for all supervisors on the financial impact of workers' comp claims
- Estimate and analyze your Experience Mod Rate (EMR) for the next rating period
- Review all open workers' comp claims

- Train HR staff in proper administration of FMLA and ADA

- Specialized safety training (lockout/tag out, confined space, HAZCOMM, OSHA 10, etc.)

- Train Supervisors in proper accident investigation techniques

- Feasibility study for captive or other alternative risk transfer/loss responsive program

- Operational risk survey

- Calculate lowest possible EMR and create plan to achieve it

- Safety training of all types

Building Trust

Hopefully, I've made it clear that trust is a critical element in your relationship with your Risk Advisor. And, if you engage in some limited scope of work to start, you'll be able to build trust in a non-pressure cooker way. Maybe it's to build a loss prevention plan. A great idea would be to send one of their loss prevention professionals into the field, visit your worksites, talk with your supervisory people, meet with your safety professionals, and that way determine how good they are and if they bring new ideas to the table. They can start saving money for you today and prove that they are worthy of your trust. And it won't cost you any money.

The same kind of project could be oriented around claim management. They could conduct a review of all your open claims to see if reserves are being set properly. They could even have an occupational physician look into a more serious open claim that involves a lot of medical bills to see if the right treatment protocols are in place. A Risk Advisor could evaluate your EMR and determine if it's calculated properly, and do some diagnostics to demonstrate what areas you should be emphasizing in your risk management plan to have the most impact on reducing your EMR. There are as many ways to engage a Risk Advisory firm as there are services on their menu. The broader their menu the more impact they're going to have on your Total Cost of Risk and the easier it will be for you to engage them on a short-term project to evaluate their capability.

Performance-Based Risk Management

Risk reversal has become a very common term used in marketing. I think risk reversal goes back to L.L. Bean. I'm not sure he was the first practitioner, but he's certainly one of the earliest. When L.L. Bean first introduced his boots, back in the nineteenth century, he told all his customers that if, at anytime during the life of that boot, something went wrong they could return it and he would either repair it or provide them with a replacement. That's what you call risk reversal. L.L. Bean was way ahead of his time in saying, "I don't want you, the customer, to take the risk. I'm going to reverse the risk. I'm going to make a guarantee."

The concept of risk reversal has never been introduced to this business of risk management, but I'm doing it here and I'm doing it now.

For our industry, I'm calling it **Performance-Based Risk Management**™ or **PBRM**™. You should have an expectation of your Risk Advisor for a certain level of performance, of both the quantity and the quality of that performance. When I say "quantity" I mean that there are certain risk-reducing services that your Risk Advisor is agreeing to provide. It could be a risk survey, a disaster recovery plan, claim management systems, and procedures driven by a software system built around your processes or loss prevention. Regardless, there's a certain quantity of risk-producing services that you should get your Risk Advisor to agree on. The second part of it is "quality." Obviously, somebody can engage in a lot of activities, or a firm can provide you with a lot of activity and loads of reports, but if the people providing those activities are ineffective or the quality isn't having the impact that you desire when you measure it, then why should you pay for it?

One of the reasons that a concept like this has never been introduced in the insurance agency brokerage world is because it's commission-based. There is no choice that a middle market contractor typically has other than to pay the premium; the premium then goes to the insurance carrier, and the insurance carrier pays the broker a commission. What I'm suggesting with Performance-Based Risk Management is that commissions are eliminated and replaced with a fee. The fee is determined by an agreement between the client firm and the Risk Advisor, and it's based on the scope of what's going to be provided as well as the capabilities and experience of the people providing it.

In the spirit of L.L. Bean, I would take that one step further and turn it into risk reversal, where all or part of that fee would be returned at the end of the engagement period (or during the course of the engagement) if the quality and quantity of services agreed upon were not delivered.

Mind you, this concept will sound like heresy to most of my peers, because we're accustomed to a world where our commissions are guaranteed, and the only fluctuation is directly connected with soft and hard market cycles. But, it's time for construction companies to demand more from the insurance community overall, particularly the Risk Advisor who protects their blind side, because contractors take risks every day. They take big risks when they bid jobs and agree to perform those jobs for an agreed-upon price. They can't be worried about their blind side, so they need a firm they can rely on to protect it. If they can't rely on that firm, if that firm can't build a plan and make improvements, then it shouldn't be paid; hence, "performance-based risk management." (*Because this is such a revolutionary concept, I've provided a broader description and analysis in Appendix A for those contractors who think this is the way to go and want to learn more*).

Profit Center versus Expense

Hearing that for the first time, most contractors will probably wrinkle their brows and wonder what I'm talking about. "Insurance can only be an expense and risk can only be an expense," is what you're probably thinking as you read this. What I'm suggesting is that if you understand risk, and control it better than your peers, then it can become a profit center because you're going to have a lower Total Cost of Risk and that will bring you a competitive advantage in the marketplace. You'll get more than your fair share of work, probably at better margins, and that whole area of risk management becomes a profit center for you instead of an expense.

Risk is an expense for those who can't control it. When it's controlled, and when you can see opportunities to control it efficiently, proactively, and consistently, it becomes a profit center.

Lots of contractors look out at the world of risk and say, "There's very little I can do to change the course of events. I'm hiring from union halls. I can't control safety on the jobsite. There are GC's out there who don't control the jobsite well. There are risks that we can't see. It is what it is and there's very little I can do about it."

Smarter construction companies are looking at this differently. Just as I have done, they look upstream at the larger companies in the construction field and can see the emphasis on safety. They can see the emphasis on wellness in the workplace and medical cost management. They can see what it means to value their workers and their workers' lives, the training that they've put into those workers, and the desire to keep them safe. There are untold numbers of ways to cut risk costs, and construction companies that want to grow and prosper in the years ahead are going to need an advisor who can bring to the table all the strategies and tactics that are currently available and help the contractor implement them.

Return on Investment Meets Return on Insurance

"Return on insurance" is a term that I've come up with to describe two things. You pay a premium to an insurance company and you pay a commission to an insurance broker, or insurance agent, or a fee to a Risk Advisor. I ask prospective clients what their return is on an insurance company; in other words, in return for the premium, what's the quality of the coverage and what are the services they're providing to make that insurance coverage work better? Similarly, I ask them to look at the insurance broker or insurance agent. Say, for example, the commission paid to them is thirty thousand dollars. If you created a percentage return, how much value is that broker providing to you in a year? Are they providing thirty thousand dollars of value? Are they providing twenty thousand? Ten? Are they providing any value at all? Or are they providing five hundred thousand dollars of value for their thirty-thousand-dollar commission because they've kept your EMR at such a low rate that you're so hugely competitive in the marketplace and earning more than your fair share of the work?

When you make decisions about your broker and your insurance carrier, you have to ask yourself, "What is my return on this investment?" Just like any other investment you make, it shouldn't be reactive, low-level service. You should look at your return on insurance and make very high demands and have very high expectations of what's delivered in return for your money.

Risk management is a competitive weapon when you use it to lower operating costs, hire better employees, and be more competitive in the marketplace. We're living in a time when less young people are going into the construction trades, when the workforce is aging, and when it becomes more and more difficult for every employer to get qualified staff. An axiom for all businesses is that whoever has the best people wins. Construction workers want to work in a safe work environment. They want to work in an environment where their employer is doing everything possible to keep them safe, and if they are injured for any reason, as valued employees, they're given the best possible medical attention so that they can get back to work as quickly as possible. Every employer's goal should be to return their employees safely home to their families at night. If you as a construction company owner have that as a goal, and you tell a Risk Advisor to build it into your plan, it's the best competitive weapon you can have. Not only will you be able to meet all the demands and expectations of your customers, the people you're building for, but you'll also have the lowest Total Cost of Risk, the best employees, and then, the best product – and it all translates to having a huge competitive advantage in the marketplace.

Alternative Risk Transfer

"Alternative Risk Transfer (ART)" is insurance jargon for the types of insurance plans that are loss sensitive or loss responsive. Most basic insurance policies are what we call "guaranteed cost." To use a simple example, if you had a worker's comp policy that is guaranteed cost, you pay the insurance carrier one hundred thousand dollars. In return for that they have to administer the worker's compensation policy and pay all the claims. If the administration of that policy is 30 percent, that means thirty thousand dollars goes for administration and they have seventy thousand left for claims. If you only have twenty thousand in claims throughout the course of that policy period, that leaves fifty thousand for the insurance company, and that's called their "underwriting profit."

In an alternative risk transfer arrangement, where the policy becomes loss responsive, you make an arrangement with the insurance company in which they share that underwriting profit with you either in whole or in part. These types of programs take various forms. The oldest one is a retrospectively rated policy. Another common form is a large deductible program, and something that's a little more exotic is a captive, or group captive mechanism. But the commonality in all of these alternative risk transfer programs is that you share in all or part of the underwriting profit (and sometimes investment income as well) and the underwriting profit is the difference between your premium and the losses that are paid.

There was a time when a construction company would have to have an annual premium of five hundred thousand dollars or more before they would be eligible for one of these programs. Through innovation, the eligibility requirements have gone lower and lower. Now there are some group captives that allow you to have as little as two hundred thousand dollars in premium and be eligible for one of these loss responsive programs. I am amazed on a regular basis by how many large construction companies have not chosen to look very seriously into these alternative risk transfer concepts.

The primary criteria that a construction company needs to look at are how well they control their own losses and their own claim experience. Obviously, you don't want to choose to participate in underwriting profit when there isn't any or when you're operating at a loss. But if you have control of your Total Cost of Risk, if loss prevention is a strong cultural value in your company and if you can have reasonable expectations of low losses going into the future, why would you want to give all the underwriting profit to another party, namely, the insurance company? Why wouldn't you want to share in all or part of that underwriting profit?

I spoke earlier about risk being a competitive weapon; this is the ultimate ace in the hole. This is the hand you can play on your competition that they can't beat. Through a combination of low Total Cost of Risk and a well-designed alternative risk transfer program, you can have the best of both worlds. Not only will you be minimizing your Total Cost of Risk, you'll be transferring as little risk as possible to the insurance company and keeping the profit for yourself.

A Risk Advisor should put every one of his or her clients on a path where the ultimate goal is to earn the underwriting profit of their premium for themselves. I strongly believe that, once you're on a path toward prevention and minimization of claims, the ultimate destination is a program where you pay the least possible premium. Alternative risk transfer programs are the most efficient way to transfer risk. That is the ultimate payoff for a safe workplace. If you're large enough – and, as I said, that threshold is now as low as two hundred thousand dollars – you can get into a loss responsive program. If you've got a safe workplace you're going to multiply the financial impact of your safety investment.

In a traditional guaranteed cost program – whether it's worker's compensation, general liability, or automobile – you're paying that full dollar. The most you can do is get into some competitive bidding arrangement, find an inept underwriter, and maybe knock 10 or 15 percent off, which isn't very much. Companies who have the safest work environments and the best loss responsive insurance programs can get 70 to 80 percent off. That's where the real money is. You don't want to be competing against contractors who have that kind of cost advantage. You want to be the leader here.

The benefits of these programs aren't just one year when the market is soft. It's not one year when you get a really inept underwriter. It's not one year when the circumstances are very competitive; it's an ongoing advantage that you have every single year, and it should be your goal. It's a one-two step: "Make my workplace the safest, and get the most efficient loss responsive program I can find." Put those two together, and nobody can beat you.

9 "I Feel Pretty, Oh So Pretty."

Every construction company presents a risk profile to an underwriter of insurance. But, this isn't something you've ever seen written on a piece of paper: "Risk Profile of Jack's Construction Company." A risk profile is made up of many different factors, which you can review below. This is a partial list, but it certainly gives you an idea of all the different areas that a good insurance underwriter is going to look at when evaluating you as a potential client.

When you or your insurance agent focuses on buying cheap insurance, you probably aren't focusing on maximizing your risk profile. Yet, maximizing your risk profile is one strategy that will guarantee you will pay less for insurance than any of your competitors. We have a unique process called Risk Profile Maximization™ (RPM™) and we use it to make you more attractive to the insurance marketplace as well as designing a plan to reduce your TCOR.

Risk Factors Contributing to or Damaging Risk Profile Maximization™

- Financial status of the company
- Experience of management team
- Safety culture
- Safety training

- Presence of full time safety director

- Annual safety improvement goals

- OSHA compliance

- Disaster preparedness

- Fleet maintenance

- Housekeeping (garage, jobsites)

- Well-written subcontract agreement

- Certificate of insurance management process in place

- Experience on jobs being bid/performed

- Medical clinic relationships

- Return to work program with coordinator

- Quick reporting of workers' comp claims

- Lost workday case rate below industry average

- Bureau of Labor and Statistics (BLS) accident and severity rates below industry average

- Accident investigation process

- Human resource compliance with state and federal laws

- Hiring practices

- Website

Your risk profile is a high impact area because it tells the insurance underwriting community what your susceptibility is to having a loss, a serious loss, or the probability that you'll have different types of losses. What kind of risks does your business present? Your risk profile is the best indication they have. There are certainly some areas in the risk profile that are much higher impact than others. As a contractor, the experience of your management team is going to be a critical factor. If you are out bidding on large, complex, multimillion-dollar construction projects, the underwriter wants to know that you can execute those jobs successfully.

Safety culture is another critical area. Is safety an afterthought in your company, or is it a value that's driven from the CEO's office right down to the frontlines? A return-to-work program is another essential element that a workers' comp underwriter is looking for. Without a return-to-work program, a work-related injury to one of your employees could be multiplied five to ten times. It could turn what should be a one- to two day absence into a highly litigated, long-term unemployment situation that costs you tens of thousands of dollars beyond what it should. A risk profile can also reveal things that might alarm an underwriter, and those are parts of your risk profile that need to be managed.

Little known fact: One of the most common areas of risk profile mismanagement is on your company website. Any business owner likes to brag and highlight their most meaningful accomplishments. On a contractor's website, this takes the form of the scariest jobs because they make construction companies look awesome and super tough. Now, the first place anyone today looks to learn about a company is their website. And, if the first thing an underwriter sees on your website is a crane extending one hundred feet into the air over a very high density work area where lots of people and material are actively engaged and the collapse of that crane would cause a disaster (as they have all over the U.S. in the last twelve months), they're going to immediately think twice about whether they want you as their next client. You might tell the underwriter that it's the only time you ever used that crane and it was ten years ago. But just the fact that you had it on your site shows them that you *have* done that type of work and you could do it again; and they know it's a potential hazard.

With something as seemingly innocuous as your website, you could eliminate an insurance company that would be the best provider of coverage and service just because you've innocently misled them about what you really do as a construction company. When you design that site, keep in mind what it says about you as a potential customer to an underwriter. A good Risk Advisor can help you do that because they don't just look at your insurance policy – they look at every single piece of your company and its identity.

Dos and Don'ts on Designing your Website

- Don't show pictures of highly hazardous work that you did one time 5-10 years ago

- Don't show pictures of unsafe acts

- Don't advertise types of work you no longer do

- Don't sound like you're a jack-of-all-trades, doing anything for anyone

- Don't hide your comments about your safety culture

- Don't build it once and then have outdated information

- Do show an impressive project list for major owner/GCs/CMs

- Do make a really big deal of safety and make sure there is a prominent link on your homepage

- Do show pictures of completed work

- Do include testimonials from major project clients

- Do feature case studies

- Do make sure all the links work and
  your site is easy to navigate

- Do list awards from trade associations, safety
  awards/recognition of any type

- Do provide a good description of current operations

- Do provide good biographical information
  on the leadership team

What You Look Like to the Underwriting Community

Earlier in the book I said it's all about the claims, and even though I've also said that past claims are not necessarily an indicator of what's going to happen in the future, it's the best information that an underwriter has to evaluate a construction company. It's why I've emphasized throughout the book that the things you have to focus on are prevention and the minimization of the claims themselves.

The most valuable service that a Risk Advisor can bring to you is that prevention and minimization of claims. Once you get beyond the claim history, then Risk Advisors look at what kind of work you do. How dangerous is it? What experience do you have? Are you going to be doing different types of work in the next twelve months than you've done in the past? Are you going to be expanding into any new geographic areas? What kind of safety culture do you have in place? Is your loss experience just because you've been lucky or do you have a great loss history because you've got a great safety culture? Many of these things underwriters can't evaluate from the paper on their desk; that's why they have someone go into the field and look directly at your operations, talk with your field people, and find out exactly how you operate at the worksite level.

Unfortunately, most insurance agents fall down when it comes to how they represent you to the marketplace. In their goal to either find a cheap underwriter or come up with the lowest possible cost, they take shortcuts in relaying the facts about your company. The difference between what a qualified Risk Advisor shows to an underwriter and the typical submission from an insurance agent is like night and day. (See Appendix B for a comparison.)

I've suggested the idea of evaluating a Risk Advisor with project work. One of the stumbling blocks to changing your risk profile is that insurance agents and brokers are usually engaged when it's time to go out to bid. They are put off until 120 days before the anniversary date of the insurance. If an insurance agent or Risk Advisor doesn't come onto the scene until 120 days prior to the anniversary date, and it's a fire drill to get all the applications into the underwriters and get it priced and get the proposal done, there isn't any time left to change the appearance of your risk factors or risk profile.

If you were really focused on this, you would hire a Risk Advisor a year or ten months in advance of your normal anniversary date, and he or she would begin to work with you in every way possible to change as many risk factors as they can. If you don't have good safety practices in place a good Risk Advisor will work with you on a project to change that He or she determines what you can do with your training and claim management, your website, and contractual risk transfer. They figure out the different elements that need fixing so that when they make a presentation to the underwriter, you look a lot prettier than you did eight months ago.

Many of the changes required for Risk Profile Maximization aren't expensive. Some of them are cosmetic. Some are just a way to take the same information you already have and present it differently or put a spin on it so it's understood more clearly by the underwriter.

You might find that some of the claims on your claim record within the last five years relate to operations that you no longer perform, and you should find a way to analyze that. Pull those out and explain to the underwriter that they should not be considered going forward because the types of operations that caused them have been discontinued. There's no specific formula for this. A good Risk Advisor will look at the high impact areas and will work hard to package you in a way that puts your best foot forward.

Loss Control Representatives

Many contractors have come to the conclusion that loss control representatives of insurance companies are not their friends, possibly because they've participated in a site visit from one or read a report resulting from one. But after the claim record, loss control representatives could be the second most important area for you to examine. Because the insurance company really doesn't trust the submission of information given to them by the insurance broker (it's incomplete, often inaccurate, and provides no analysis) they send their own representative out to talk with representatives of your company and to actually go look in the field to see if what they see on paper is really the truth.

Once again, this is something that has a critical impact on the cost of your insurance. It's a chance for you to evaluate whether or not the safety representative of the insurance company is competent, understands construction, is going to cooperate with you, is going to act with a heavy hand, or understands what you're up against every day and is going to be understanding of the human side of implementing safety practices.

Keep in mind that even though safety is often delegated to the safety professional, no one believes in the company's mission and understands the company as well as the owner/CEO. This is your opportunity to make your case to an insurance company about why you'd be a great client. Don't delegate this task to anyone else. Your safety director is probably not going to communicate your company's mission and values the way you would; and even though there are probably many aspects of the safety program that you can't discuss at the same level of detail as your safety director, you can give the overriding view and really sell the company on your commitment as the owner. There's nothing an insurance company wants to hear more than that the CEO/owner/executive team is committed to safety and why they're committed to safety, and you're the one who can speak that message best.

10 Why God Invented A.M. Best

Rating Organization

A.M. Best is the most widely recognized insurance company rating organization. Insurance companies are financial institutions, and just like S&P and Moody's rate bonds, A.M. Best rates insurance companies on their claim paying ability. A.M. Best employs top economists, actuaries, and analysts that pore over insurance company financial statements and try to make some determination of the financial soundness of insurance companies. As a contractor, you've probably read that a certain Best rating is required of the insurance companies you provide on a certificate of insurance. Usually that rating is "A" or better. A.M. Best is like other rating organizations in that they have an alphabetic range of ratings from A to F where A++ is the highest and F is the lowest.

Disclaimer: I'm writing this book in the beginning of 2009 when we've just seen every rating organization miss the mark in terms of what they should be telling consumers of financial products. No one knew Lehman Brothers and Bear Stearns were going to go out of business. AIG had been the strongest insurance company on the planet for eons; they were rated A+. A few months ago they were bailed out to the tune of eighty five billion dollars by the federal government. You could be cynical and say, "What good are these rating organizations if they tell me an insurance company is A+ one day and they're in receivership with the federal government the next day?" The answer is that at the moment it's all we've got, and I'm hopeful that many improvements will be made in the systems and processes of these rating organizations used to evaluate financial stability in the future. I hope the industry will learn from its mistake, but that remains to be seen.

So, going with all we've got (this current rating system), it's imperative that you pay attention and work with top rated insurance companies. First, you'll continue to see job specifications which require using insurance companies with an A rating or better. You or your Risk Advisor will have to comply with that requirement. If you get midterm in a contract and your insurance company loses their rating, falling below the required level, you're going to have to replace them or you're going to be in noncompliance with your contract.

And, second, you'll want to stay compliant for more than simply the legality of staying compliant. When we talk about protecting your blind side, we're talking about a long-term play – this is where the football analogy ends. In football, the play is made in an instant and then you can evaluate the outcome. But, if you have a serious liability claim that happens today, you might not end up in a courtroom for two, three, four, or five years from now, depending on the type of incident and the complexity of the litigation. It could be appealed, it could be a long jury trial, and it could even be longer before a jury finally reaches a decision. So, you need to pick a top rated insurance carrier that will stand the test of time, that will be able to pay that claim when it finally comes due, especially if it's of a catastrophic nature and for millions and millions of dollars.

Worst-case scenario, you protect your blind side with expensive insurance coverage and the best level of risk advice that money can buy, and then the insurance carrier is not there to pay the claim. Then it's all been for naught. This is very serious business and it's something that you really need to pay attention to. You can be proactive by consistently focusing on the quality of the coverage and not the cost of the premium. If you've been paying attention, the cost of the premium will take care of itself once you have your TCOR ducks in a row.

Is local (insurance) better?

Another choice you have to make when deciding who your insurance company will be is whether it's a local, regional, national, or even international carrier. There are pluses and minuses to all of them, and it probably makes sense to understand the major differences. If you're dealing in a local market, limited to one state or one region, a local carrier might be a good choice – and I say "might" because they're not all created equal. Some local regional carriers are very impressive in what they bring to the table: they have a high customer focus, they're financially sound, they're not taking big risks, and they understand their customers very well. If you operate outside of a local area, though, they might not be able to provide the services that you need. If you are in a multi-state trade area and all their claim offices are located in one state, they're not going to understand the legal issues of those other states you operate in. They might not have local loss prevention people on the ground in those states, so their service will suffer.

Look at the scope of your operations. If your operations are beyond a local region you really have to consider a national carrier. National carriers can be the biggest and the strongest, but I think as we've learned lately that the bigger they are the harder they fall. AIG until recently has always been one of the most profitable insurance carriers in the world, and here they are needing a bailout form the Feds. From that respect, a smaller regional carrier that isn't loaded up with a bunch of risky investments might have been a safer bet than AIG.

I guess we can look at AIG and say, "Who cares what their financial rating is because the government stepped in and they're going to help them meet their obligations." But there are many insurance companies that don't have the scope and scale of AIG, and they wouldn't be rescued.

Always Pick a Specialist

It seems that specialization is the order of the day in our national and international economy. Whether it's in construction or medicine or publishing or law or virtually any other aspect of the economy, it seems that all of us are being forced into narrower and narrower specialties. A construction company looking for advice and protection in the area of risk management needs to find specialists in everything they do.

Let's start with the insurance company. Some insurance companies are big generalists in that they offer everything to everyone; they haven't developed a sophisticated level of specialization. There are other very large insurance companies who offer a diverse menu but are specialists in every one of them.

You have to find an insurance carrier who specializes in construction, and you want them to specialize in a big way. What I mean by that is this: the underwriter has to be a specialist in construction; construction should be all he or she works on every day. The safety people need to be specialists in construction. The last thing you want is a visit to one of your job sites by a loss prevention specialist who spent the morning in a manufacturing facility, who's coming to see you and then going to look at an apartment complex. You need someone who understands what you face in the construction industry, and who can offer practical suggestions.

The same is true on the claim side. Liability claims related to construction are a very complex form of litigation. You don't want a claims specialist who spends most of his or her time evaluating slips and falls or car accidents to be the one evaluating a multi-party contract dispute involving bodily injury to a pedestrian on a construction site. You also need to know that the lawyers, whether they're internal or external to the insurance company, are also specialists in construction. So, every piece of what the insurance company provides requires specialization for construction.

This is just as important for your Risk Advisor or insurance broker. Just like insurance companies, many insurance brokerage firms are generalists. Their individual people are generalists throughout the day dealing with all kinds of different clients and some of them are generalists in terms of how they present themselves to the marketplace, but internally they're broken down into specialty divisions. You need to find a broker or Risk Advisor who understands construction and who, for the most part, has a team advising you that advises contractors exclusively.

As time evolves, the insurance industry will probably get even more specialized than what I'm suggesting. I recently saw an advertisement for an orthopedic surgery firm. They have a specialist for the hand, another one for the elbow, another for the hip, another for the knee, one for the shoulder, and one for the spine. There was a time when an orthopedic surgeon dealt with all the bones in the body. Maybe five years from now all one subspecialty will deal with is the thumb, and maybe five years from then it's going to be left thumbs versus right thumbs. Who knows how ridiculous or how far this will go, but I can certainly see a day when specialists for construction risk and insurance management will get more specialized than they are today. As soon as possible, you want your team and your advisors to know everything about what you do and the ins and outs of your industry – and only your industry.

11 Riding the Cycle: "Give It Time."

The insurance industry, like other industries, goes through a soft to hard pricing cycle. It's probably no different than airlines: when bookings are low, prices get discounted, then they start losing too much money, increase prices, causing, once again, less people to travel. They discount once again and the cycle is complete.

Insurance has its own cycle system, and while they're different today than they were a decade ago, they still exist and it's important to understand what they mean. The most common mistake made by contractors is going out to bid every year when the market cycle is soft, thinking they are going to get better deals. It's better to choose a Risk Advisor who is fee-based and not commission-based, and who has no incentive whatsoever other than to maximize your risk profile, package you, and present you to the marketplace in the best possible way. You want someone whose mission is to reduce your Total Cost of Risk and get you a loss sensitive program that's going to reward you for all the hard work that you've done.

If you do all that, market cycles become a lot less important. As you can see in the illustration on the following page, you want to ride the lower curve during these market cycles. This way it won't matter what part of the cycle you're in because you're always going to be paying less money than most of your competitors. When you go out to bid during soft market cycles, it's easy to think that the insurance agent is the hero saving you a lot of money when it's really just the crazy market or the inept underwriter that has saved you money.

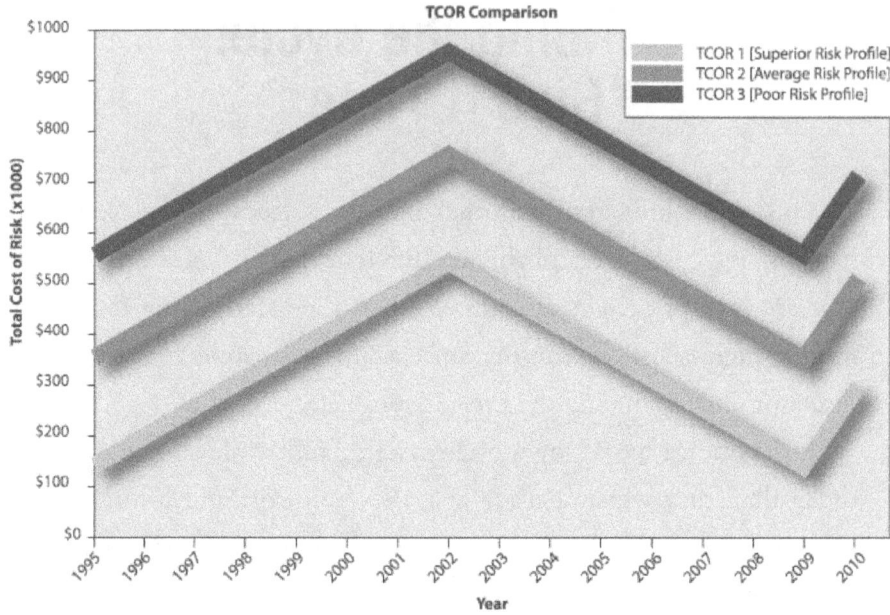

Also, the lower your premium gets, the more likely it will become a case of being careful what you wish for. The truth is, if you're not focusing on your claims and prevention and the Total Cost of Risk, your claims remain at the same level but your premium keeps going lower and lower, which means that every year in that soft market cycle you're going to be a less profitable account to your insurance company. If you stay focused on your minimizing your TCOR as your premium goes down then you're going to be okay. That's the way you want it to be. But if all you do is focus all your attention on paying less for your insurance, sooner or later you're going to be upside down and your losses are going to exceed your premium. The market is going to turn and you're going to be an undesirable account for an insurance company with a lousy risk profile. You might go from spending two hundred thousand dollars one year to seven hundred thousand the next (that kind of extreme shift is not unusual at all). If that's compounded by a soft economic cycle like we're in right now in the end of 2008 and the beginning of 2009, all of a sudden you could be really upside down. That is a perfect example of getting hit on your blind side. It could be "game over" for your company.

By this time you probably can see that for a Risk Advisor to really do their job, they must be in it with you for the long term. A Risk Advisor's job is a much higher level of commitment than a typical insurance agent. They really need to understand your business. They're going to commit a lot of resources to understanding your business and helping you with it, and they want to know that you're committed for the long term as well. They need to know that you're not going to go out to bid every year and flip the relationship because someone has a cheaper price. This is a two-sided commitment with a heavy dose of dedication.

If you look at other professionals that advise you, CPAs, lawyers, etc., you want those types of professionals to get to know you; you're going to have multi-year relationships with them in almost every case. It needs to be the same with a Risk Advisor. If someone's going to protect your blind side, they must really be up to their eyeballs in your company and know it very well. You might think that shopping every year will get you the best deal, and as I've said earlier, that's the fool's gold. It might look that way but slowly and surely it's going to backfire.

Final Thoughts

The purpose of this book is threefold. I'm asking you to reconsider how you view risk, how you buy insurance, and how you choose the advisor who helps you with that. I've introduced some new concepts and some new methodologies. What I believe is that you need to protect your blind side. You, your family, and your employees have all built an incredible company that provides a valuable service to your community, whether it's to your customers, employees, or vendors; they're all stakeholders in your success.

If you shift your emphasis from buying cheap insurance to attaining the lowest possible Total Cost of Risk, it will put you on a new path to success. When you maximize your risk profile, reduce your claims, and develop a culture of safety, you're going to be able to take advantage of the most efficient loss responsive insurance plan. Once you're able to allocate less money to the protection that you've traditionally purchased, you'll then be able to allocate more money to the prevention of those catastrophic events that can hit you in that blind side.

The goal here is to minimize your losses, buy more efficient coverage and broaden your protection so that nothing can ever put your most valuable asset at risk again.

The path to that end is finding, testing, hiring, and building a relationship with a Risk Advisor, someone who will proactively protect your company and earn their money in direct correlation to the performance of their work for you and the comprehensive protection plan they have built around your most valuable asset.

The moral of this story is that risk management will sustain, protect, and grow your company – cheap insurance will only bite you on the ass.

Appendix A

The advent of performance-based professional fees

For several years, my firm has been talking to our clients and prospects about what I like to call their "Return on Insurance" or ROI. When comparing our value proposition to that of another broker, we ask how much value is provided for the commission dollars spent. Unfortunately, the way insurance transactions are typically structured, the insurance company pays the agent or broker in the form of a commission on the products sold. Because of the antiquated anti-rebate laws in most states, the commission amount can't be changed to reflect the level of value provided. Nor can a broker give any back at the end of the year if they didn't deliver as promised. For you, the contractor client, this means that regardless of the quantity or quality of service provided by your broker or Risk Advisor, you pay them the same amount of money.

Why should a professional services relationship between two parties be based on product-based commissions? This only makes sense if the exclusive focus of the relationship is based on the products. Unfortunately, this is the case 95% of the time because most agents and brokers don't focus on Total Cost of Risk but on selling cheap insurance. Commission-based compensation makes sense in a world of commodity-type sales where consumer choice is based strictly on price. But as I've emphasized in this book, this should never be the case for a middle market construction company worth many millions of dollars.

So why hasn't the system changed? Why do insurance agents and brokers sit complacently while the business models of every other industry on earth are changing by the day if not by the minute? In the thirty plus years I have been practicing, nothing in the business model of insurance agents and brokers has changed. Why have they been protected?

Look at how new business models have affected the airline industry. The winners are JetBlue and Southwest, the innovators. All the others have gone bankrupt or have one foot in the grave. Look at the American auto industry. They are hopelessly behind relative to the Japanese, German, and Korean manufacturers. Now it's too late to change. They will never recover their lost market share and can only hope to survive in some new, smaller, less significant form. Even in computers there has been major disruption. IBM invented the PC but couldn't compete and sold the division to Lenovo. Dell was dominant with their direct sales model and efficient manufacturing. Now HP has eclipsed them by outsourcing to China. Lower cost and more value to the consumer has been the result.

If you compare insurance agents and brokers to other commission-based service professionals, they are the only ones whose business model hasn't evolved. Disruptive change, for the right reasons, has altered the business model of other commission-based professions. Stockbrokers were the first to face change and that was in the 70's when Charles Schwab came on the scene as the first discount broker. Real estate agents lost their exclusive franchise with the concept of "for sale by owner" followed by negotiated commissions. Travel agents have been completely disintermediated by Internet services like Orbitz, Priceline, and Expedia. If you still use a travel agent, you don't pay them a commission based on the cost of your travel, you pay them a fee. Similarly, CPA firms negotiate fees on an individual basis with every client. And look at doctors. Even though they are the most highly trained of all the professions, managed care has changed their compensation forever.

The only types of professional service firms that haven't changed their business models are law firms and insurance brokers. Law firms have stuck to the billable hour and insurance firms have hung onto their commissions.

Back to my question. Why hasn't the compensation system for insurance agents and brokers changed? For the same reasons it hasn't changed in law firms. The leaders of these firms are highly compensated and don't want to change. It means they would have to become a lot better or a lot more efficient or both. They should hear the challenge from the marketplace shouting louder and louder, "change or die, change or die" but they turn a blind eye and deaf ear. They don't look in the mirror and realize they are staring at a dinosaur (or an auto industry executive). They think if they can hang on just a few more years it will all go away or they will be retired.

Change is not completely absent from the legal profession but it was well hidden until just recently. At the very bleeding edge, some renegades from large law firms have created a new business model and at least for certain types of work and projects (primarily litigation), it does away with the billable hour. I first read of this concept as it is practiced at a boutique litigation firm in Chicago called Valorem Law Group, whose slogan is, "The billable hour is dead." The other law firm I have recently heard is experimenting in this area is Summit Law in Seattle, Washington. But probably the most significant endorsement of this concept came in an article written by Evan R. Chesler, the presiding partner at white shoe NYC law firm Cravath, Swain & Moore, LLP, in the January 2009 edition of *Forbes* magazine. About the billable hour Mr. Chesler says,

> *"The billable hour makes no sense, not even for lawyers. If you are successful and win a case early on, you put yourself out of work. If you get bogged down in a land war in Asia, you make more money. That is frankly nuts."*

Patrick Lamb, a partner at Valorem Law, blogs on this subject. In a recent post he says,

> *"For outside counsel at firms that raise rates this year, my advice is to immediately go to your phone, pick up the receiver and dial 1-800-PSYCHIATRICHELP and follow the instructions you receive."*

If looked at purely from the concept of placing the insurance in the traditional role of insurance agents and brokers, they make more money when their clients' premium goes up than when it goes down. How ridiculous is that? Talk about misaligned incentives. If Risk Advisors and brokers are going to be paid less and less as they reduce their clients' Total Cost of Risk, why would they even try? No wonder the insurance industry has a trust issue with its clients and no wonder agents and brokers are clinging to a broken compensation system. Just as in the law, it makes no sense for either party.

There are two pieces to this compensation problem. The first is that all insurance agents and brokers are paid the same standard commissions on the policies they sell. Regardless of wisdom, skill, or value proposition, they're all paid the same. If I use a law firm/CPA firm analogy, why would you pay an associate the same as a senior partner? Why would you pay a small two-person CPA firm the same fee as a large regional firm with more resources, broader capabilities, and a more seasoned professional staff? The second issue is one of performance. Why do middle market businesses agree to pay a fixed commission to an insurance agent regardless of how well they do their job? If your Risk Advisor's ultimate goal is to reduce your TCOR and as a result your insurance costs go down, that means that the better they do their job the less money they make. They make *more* money when *you* suffer and your insurance premiums go up! Is there something wrong with this picture? Let me repeat, just to be clear. The more you, the consumer, pays for insurance, the more you pay your broker. ***In other words, the worse it is for you, the better it is for them***. And this has been an acceptable practice for how long? How about forever.

The revolutionary concept I propose to fix this problem is called ***Performance-Based Risk Management*™ (PBRM™)**. Here's how it works. To solve the compensation piece, we eliminate commissions to the greatest extent possible. In some cases it's impossible to eliminate them entirely because of state laws mandating that commissions be paid on certain lines of coverage (I'm ready to join you if you'd like to lobby to change this). Also, we Risk Advisors have to go through wholesale brokers on some insurance products and we can't force them to do business our way. This is full disclosure on how the system works but you still get exactly what you want which is a fee commensurate with the work performed. We (Risk Advisor and client) will mutually agree on what the fee will be. If we earn any commissions they will be deducted from the fee so the most you will pay is the agreed upon fee. Full transparency. No surprises. No extras. Just the agreed upon fee.

I don't want to create the impression that just because a Risk Advisor is compensated on a fee that the fee will be less than an equivalent commission. The fee will be appropriate, no more and no less. If your risk management program is a disaster and your losses are out of control and you need immediate assistance, then the fee will reflect that. Quality of services, quantity of service, experience of staff, and breadth of services required will all be considered when establishing the fee. This is not a discount program. The purpose of PBRM is to match an appropriate level of compensation with an appropriate level of service or an appropriate solution.

However, it's the second component of PBRM™, which makes it interesting and revolutionary. We already have many clients who pay us a fee arrangement as I've just described. This has been a common practice with Fortune 1000 and upper middle market companies and their brokers for years. To my knowledge, though, no one has this Risk Reversal piece. I learned this from the Valorem Law Group. They call it the "Value Adjustment Line." It works like this. We establish an annual fee or a project fee. We then bill the fee monthly. If you feel we have delivered more value or *less* value for the fee being charged then you are completely free to adjust the invoice and pay a greater or lesser amount (up to and including nothing). How about that for putting our money where our mouth is? I'm sure L.L. Bean would be proud. Here's how Valorem Law describes it on their website:

> *"We Provide Value or You Adjust Our Fees. On each bill, you have the right to make any adjustment to our proposed fee that you feel is needed. We provide value or you adjust the bill, it's that simple. We do this to give you the ultimate check on our unwavering commitment to client service, and to eliminate the concern that our level of service will wane once the work we've performed exceeds a given flat rate or capped fee allotment. Some have said that the Value Adjustment Line is extremely risky. We agree. If we aren't willing to risk our own fees on our service, do you really want us advocating for you? Is this a risky approach? Yes. But it's a risk we're willing to take because of our deep belief in the quality of what we provide. How can we ask our clients to believe in our commitment if we don't believe in it ourselves?"*

I feel the exact same way. I invite you to learn more about PBRM™ and our revolutionary approach to risk management by calling me directly at 860.761.7201 or visiting our websites **www.constructionra.com** or **www.totalcostofrisk.com**.

Appendix B

Submission comparison between insurance agent and Risk Advisor
"Who would you rather have buying your insurance for you?"

	Insurance Agent	Risk Advisor
Applications	X	X
Loss runs (5 years)	X	X
Description of legal entities		X
Pictures of physical property		X
Pictures of jobsites		X
Excel spreadsheet of exposure basis		X
Narrative description of company history, operations, growth strategy, etc.		X
Organizational chart and bios of key people		X
Analysis of large losses		X
Copy of safety manual		X
Copy of employee manual		X
Copies of OSHA logs		X
Description of light duty program		X
Safety training schedule		X
Disaster plan		X
WIP (work in process)		X
Major job list (5 years)		X
Subcontract agreement		X
Sample owner/GC agreement		X
Analysis of workers' comp claims by injury type, cause, time, etc.		X
BLS accident and severity rates		X
Certificate of insurance management procedures		X

Accident investigation process		X
Hiring procedures		X
Alcohol and drug testing procedures		X
Historical 5-year summary of losses, premium, and exposure		X
Experience Mod Rate (EMR) worksheets for most recent 3 years		X
DAS/DOT certifications from state		X
Financials		X

Contact Information

Robert G. Phelan, ARM, CRIS
Construction Risk Advisors
Chairman & CEO
126 South Main Street
PO Box 1127
Torrington, CT 06790

800-252-9864 (Office)
860-761-7201 (Office Direct)
860-751-9607 (Mobile)

Email: bphelan@constructionriskadvisors.com

LinkedIn profile: http://linkedin.com/in/bobphelan

Follow me on Twitter @constructionra or @totalcostofrisk

Robert Phelan, ARM, CRIS
Chairman & CEO
Litchfield Insurance Group
&
Construction Risk Advisors

Robert Phelan is a thirty year veteran of the insurance industry. Having worked for major brokerage firms throughout New England, Bob is currently Chairman & CEO of Litchfield Insurance Group and its subsidiary, Construction Risk Advisors.

In his capacity as CEO, Bob has led his company to national prominence in providing a wide range of value-added services to its clients. Litchfield was named by *Rough Notes* Magazine as Marketing *Agency of the Month* for May 2001 and then received the ultimate recognition of *Agency of the Year 2001* by the same publication.

In January '05, after conducting a nationwide search, The National Alliance Research Academy recognized Mr. Phelan as one of "*The 25 Most Innovative Agents in America*". The profiles of all 25 were published in March '05 in a book by the same name. Bob has just published his first book, ***Broke – The Broken Contractor's Insurance System and How to Fix It*** (4/09) and has recently written a chapter on *Innovation* in a book titled "**Secrets of Peak Performers**" which will also be published in April '09. His co-authors are nationally recognized marketing experts Dan Kennedy and Bill Glaser and executive performance coach Lee Milteer.

Bob has a BA from Tufts University, holds the professional designations, Construction Risk and Insurance Specialist (CRIS), Associate in Risk Management (ARM) as well as Certified WorkComp Advisor (CWCA). Mr. Phelan is presently enrolled in the Associate in Captive

Insurance (ACI) program sponsored by the International Center for Captive Insurance Education (ICCIE). He will complete the program and attain the designation in mid '09. In addition, he is a distinguished graduate of The Buckley School of Public Speaking. Mr. Phelan has taught insurance at the college level.

Bob is a member of the Editorial Advisory Board of *Rough Notes* Magazine, the largest circulation trade journal dedicated to the insurance brokerage business. In February of '09 Mr. Phelan was recognized by *Risk and Insurance Magazine* as one of six *Power Brokers* in the U.S. in the **Construction** category. This is an annual listing of the most influential commercial insurance brokers in the U.S. categorized in 25 industry practice groups.

Mr. Phelan was invited to join Travelers Insurance Company's *National Construction Advisory Council* in January '09. He is former President of Captiva, Ltd., a Bermuda-based captive insurance company and has spoken extensively on the captive concept. Bob has also spoken throughout the U.S. and Canada on the topic, "Differentiation through Value-Added Services." Mr. Phelan is a former member of the Advisory Board of Sitkins International, the most prestigious insurance broker network with members throughout North and South America.

TreeNeutral

Advantage Media Group is proud to be a part of the Tree Neutral™ program. Tree Neutral offsets the number of trees consumed in the production and printing of this book by taking proactive steps such as planting trees in direct proportion to the number of trees used to print books. To learn more about Tree Neutral, please visit **www. treeneutral.com**. To learn more about Advantage Media Group's commitment to being a responsible steward of the environment, please visit **www.advantagefamily.com/green**

BROKE is available in bulk quantities at special discounts for corporate, institutional, and educational purposes. To learn more about the special programs Advantage Media Group offers, please visit **www.KaizenUniversity.com** or call 1.866.775.1696.

Advantage Media Group is a leading publisher of business, motivation, and self-help authors. Do you have a manuscript or book idea that you would like to have considered for publication? Please visit **www.amgbook.com**

www.ingramcontent.com/pod-product-compliance
Lightning Source LLC
Chambersburg PA
CBHW051548170526
45165CB00002B/927